NANOCARRIER TECHNOLOGIES

Nanocarrier Technologies:
Frontiers of Nanotherapy

Edited by

M. Reza Mozafari
Massey University, New Zealand

 Springer

A C.I.P. Catalogue record for this book is available from the Library of Congress.

ISBN-10 1-4020-5040-2 (HB)
ISBN-13 978-1-4020-5040-4 (HB)
ISBN-10 1-4020-5041-0 (e-book)
ISBN-13 978-1-4020-5041-1 (e-book)

Published by Springer,
P.O. Box 17, 3300 AA Dordrecht, The Netherlands.

www.springer.com

Printed on acid-free paper

Printed in the Netherlands

Dedication

This book is dedicated to my parents Ali and Susan

Contents

Contributing Authors

Ajay AWATI
Riddet Centre, Massey University, Private bag 11222, Palmerston North, New Zealand

E. Turker BARAN
3B's Research Group-Biomaterials, Biodegradables and Biomimetics, University of Minho, Campus de Gualtar, 4710-057, Braga, Portugal; and: *Department of Polymer Engineering, Univiversity of Minho, Campus de Azurem, 4800-058, Guimaraes, Portugal; e-mail: turker.baran@dep.uminho.pt*

Wangxue CHEN
Institute for Biological Sciences, National Research Council of Canada, Ottawa, Ontario, Canada K1A 0R6

Costas DEMETZOS
Department of Pharmaceutical Technology, School of Pharmacy, Panepistimiopolis, Zografou 15771, University of Athens, Athens, Greece. Tel: +30210 7274596; Fax: +30210 7274027. e-mail: demetzos@pharm.uoa.gr

Ashraf S. ELKADY
Egyptian Atomic Energy Authority (EAEA), NRC, P.O. Box 13759, Cairo, Egypt; e-mail: ashraf_elkady@yahoo.com

John FLANAGAN
Riddet Centre, Massey University, Private bag 11222, Palmerston North, New Zealand

Konstantinos GARDIKIS
   Dept. of Pharm. Technology, School of Pharmacy, University of Athens, Athens, Greece

Michal GRZYBEK
   University of Wroclaw, Institute of Biochemistry and Molecular Biology, Wroclaw; Poland

Sophia HATZIANTONIOU
   Dept. of Pharm. Technology, School of Pharmacy, University of Athens, Athens, Greece

Kang Moo HUH
   Chungnam National University, Dept. of Polymer Science and Engineering, Yuseong-gu Gung-dong 220, Daejeon, 305-764, South Korea

Arkadiusz KOZUBEK
   Institute of Biochemistry and Molecular Biology, University of Wroclaw, Przybyszewskiego 63/77, 51-148 Wroclaw, Poland

Marek LANGNER
   Institute of Physics, Wroclaw Technical University, Wyb. Wyspianskiego 27, 50-370 Wroclaw, Poland

Sang Cheon LEE
   Korea Institute of Ceramic Engineering and Technology, 233-5 Gasan-dong, Guemcheon-gu, Seoul 153-801, South Korea

Paul J. MOUGHAN
   Riddet Centre, Massey University, Private bag 11222, Palmerston North, New Zealand

M. Reza MOZAFARI
   Riddet Centre, Massey University, Private Bag 11 222, Palmerston North, New Zealand; email: mrmozafari@hotmail.com, mozafarimr@yahoo.com

Abdelwahab OMRI
   The Novel Drug and Vaccine Delivery Systems Facility, Department of Chemistry and Biochemistry, Laurentian University, Sudbury, Ontario, P3E 2C6, Canada

Tooru OOYA

School of Materials Science, Japan Advanced Institute of Science and Technology, 1-1 Asahidai, Tatsunokuchi, Ishikawa, 923-1292, Japan

Kinam PARK

School of Pharmacy, Purdue University, West Lafayette, IN 47907-1336, USA; Tel: (765) 49451-7759; FAX: (765) 496-1903; e-mail: kpark@purdue.edu

Girishchandra B. PATEL

Institute for Biological Sciences, National Research Council of Canada, 100 Sussex Drive, Ottawa, Ontario, Canada K1A 0R6; Phone: (613) 990-0831; FAX: (613) 941-1327; e-mail: girish.patel@nrc-cnrc.gc.ca

Thomas RADES

School of Pharmacy, University of Otago, P.O. Box 56, Dunedin, New Zealand

Rui L. REIS

3B's Research Group-Biomaterials, Biodegradables and Biomimetics, University of Minho, Campus de Gualtar, 4710-057, Braga, Portugal; and: *Department of Polymer Engineering, Univiversity of Minho, Campus de Azurem, 4800-058, Guimaraes, Portugal*

Anne SAUPE

School of Pharmacy, University of Otago, P.O. Box 56, Dunedin, New Zealand

Aleksander F. SIKORSKI

University of Wroclaw, Institute of Biochemistry and Molecular Biology, and: *University of Zielona Gora, Institute of Biotechnology, Monte Cassino 21b, 65-651 Zielona Gora, Poland; email: afsbc@ibmb.uni.wroc.pl*

Harjinder SINGH

Riddet Centre, Massey University, Private bag 11222, Palmerston North, New Zealand

Katarzyna STEBELSKA

Academic Centre for Biotechnology of Lipid Aggregates, Przybyszewskiego 63-77, 51-148 Wroclaw; Poland

Zacharias E. SUNTRES

Medical Sciences Division, Northern Ontario School of Medicine, Lakehead University, Thunder Bay, Ontario, P7B 5E1; Canada

Kyriakos VIRAS
   Dept. of Chemistry, Lab. of Physical Chemistry, University of Athens, Athens, Greece

Matthias WAGNER
   Mettler-Toledo GmbH, Business Unit Analytical, Sonnenbergstrasse 74, CH-8603, Schwerzenbach, Switzerland

Paulina WYROZUMSKA
   University of Wroclaw, Institute of Biochemistry and Molecular Biology, and: *Academic Centre for Biotechnology of Lipid Aggregates, Przybyszewskiego 63-77, 51-148 Wroclaw; Poland*

Renat I. ZHDANOV
   Institute of General pathology and Pathophysiology, Russian Academy of medical Sciences, 125315 Moscow, Russian Federation

Preface

Nanotherapy in a broad sense is the application of nano-scale technologies, including nano-encapsulation (or nanocarrier) systems, to increase life standards of humans and animals. Advances in nanotherapy and nanopharmaceutical technology have made it possible to target human and animal diseases precisely at their source whilst minimising side effects. These modern scientific fields have encompassed the application of nanoliposomes, niosomes, archaeosomes, nanoparticles, micelles and other carrier systems. Despite the fact that these carrier systems have been recently introduced, intensive scientific investigation has dramatically improved the knowledge base in the field. This, in turn, has paved the way for their potential clinical applications as evidenced by some pharmaceutical, cosmetic and food products, which have already been approved by the regulatory authorities for human use. Nanocarrier systems not only provide protection and controlled release of the incorporated material, but also have the potential to deliver their load precisely to the required site in the body - hence removing the need for consumption of large quantities of drugs and other bioactive agents. The list of materials that can be incorporated to the carrier systems is very exhaustive ranging from amino acid or nucleic acid-based therapeutics to tissue regeneration and weight-loss formulations. Results of numerous pre-clinical and clinical studies show that bioactives, such as antineoplastic agents, encapsulated in carrier systems exhibit reduced toxicities and enhanced efficacies. Given the advantages that nano-carrier systems provide, compared to conventional pharmaceuticals, it is likely that the number of formulations containing these systems approved for human and animal use will increase in the near future.

   Nanoencapsulation systems vary in terms of ingredients, rigidity, stability, release properties and ability to incorporate materials with different solubilities. The choice of which carrier system to use depends on the characteristics (size, solubility, charge, etc.) of the bioactive agent to be encapsulated as well as the intended application. In this regard, a general knowledge on currently employed carrier systems will be very useful. Towards this end the present book intends to collate recent advances in the field of nano-encapsulation technologies and their application for treating some of the contemporary human health challenges. Expert reviews on the major types of carrier systems and their role in areas such as cancer, gene therapy, oxidative stress and food nanotechnology provide excellent opportunity to gain insight into the field. Referencing and citation in each chapter was left to the choice of authors to whom I am very much indebted for responding to my invitation to contribute. I am also grateful to Springer for publication of this book.

   Thanks to the high quality of the chapters, backed by several years expertise of the authors, I have no doubt regarding the usefulness of this book as an ideal source for researchers, lecturers and scientists in the field of nanotherapy. Hope that future brings more health and peace for our children.

                                                M. Reza Mozafari, PhD
                                                    February 2006
                           Massey University, Palmerston North, New Zealand

Foreword

Although the prefix "nano" entered impetuously into our lives in the latter decade of the 20th century, penetrating habitual frames/barriers and notions, just a few years ago nobody had even heard about "nanotherapy". The nanoworld appeared suddenly before our eyes, which had become accustomed to a microworld. Some people, maybe even scientists, believe naively that it is enough just to add the "nano" prefix instead of the common "micro", to get a new quality. The present volume is not one of those. It is devoted to new and splendid possibilities which the world of nanoparticles grants to people and their beauty and health. It is accepted that the "nanoworld" starts for us from the 100nm size (and less), which is reflected excellently in the present book, being one of the first in the field of nanotherapy. It is not an exhaustive edition (although it is always better to finalize a not perfect volume, than never finish a perfect one), but it covers the most state-of-art nanotechnologies and materials used for nanotherapy. Some topics such as clinical implications are not covered in this volume, but can be found elsewhere.

The Editor, Dr. M. Reza Mozafari, being a young talented scientist, appears already to be an experienced editor. His excellent credentials permit that. He succeeded to choose and collect a number of excellent chapters from bright scientists and top laboratories around the world. The book will without any doubt perfectly fit into the "nano" literature shaping the field.

I am sure that the benevolent efforts of the authors and the editor constructively serve the international scientific community and all of mankind.

Renat Zhdanov, DSc, Professor of Biophysics
Russian Academy of Medical Sciences, Moscow, Russian Federation
March 2006

Acknowledgments

I would like to thank all contributing authors for their dedication and hard work in producing their excellent articles. My special thanks to my wife and children for their support and patience during the hard task of editting this book.

M. Reza Mozafari, PhD
February 2006
Massey University, Palmerston North, New Zealand

Chapter 1

BIOACTIVE ENTRAPMENT AND TARGETING USING NANOCARRIER TECHNOLOGIES: AN INTRODUCTION

M. Reza Mozafari

Riddet Centre, Massey University, Private Bag 11 222, Palmerston North, New Zealand

Abstract: Optimal nanotherapy requires the bioactive or therapeutic molecule to be protected from degradation and reach its target cell and intracellular location. Because of their unique properties, nanocarrier systems enhance the performance of bioactives by improving their solubility and bioavailability, *in vitro* and *in vivo* stability, and preventing their unwanted interactions with other molecules. Nanocarrier technologies used for the entrapment/encapsulation and selective targeting of bioactives can be broadly categorised as polymer, lipid and surfactant-based systems. Cell-specific targeting of carrier systems is a prerequisite to attain the concentration of bioactives required for therapeutic efficacy in the target tissue while minimising adverse effects on other parts of the body. Methods of entrapment of bioactive agents in nanocarrier systems and their targeting strategies are described in this chapter.

Key words: Bioactive agents, nanocarriers, nanotherapy, targeting.

1.   INTRODUCTION

The protection of bioactive agents, including drugs, vaccines, nutrients and cosmetics, from degradation and inactivation has been investigated extensively using microencapsulation systems. However, to provide targeted controlled release is a key functionality that can be provided much more efficiently by employing nanocarrier technologies. Advancements in nanoscience and technology have made it possible to manufacture and analyse sub-micrometric bioactive carriers on a routine basis. The delivery of bioactives to various sites within the body – as well as in non-living systems – and their release

1

M.R. Mozafari (ed.), Nanocarrier Technologies: Frontiers of Nanotherapy, 1–16.
© 2006 *Springer. Printed in the Netherlands.*

behaviour is directly affected by particle size. Compared to micrometer size carriers, nanocarriers provide more surface area and have the potential to increase solubility, enhance bioavailability, improve controlled release and enable precision targeting of the entrapped compounds to a greater extent. As a consequence of improved stability and targeting, the amount of material required to exert a specific effect when encapsulated or incorporated to nanocarriers is much less than the amount required when unencapsulated. This is particularly useful when dealing with expensive bioactive materials. A timely and targeted release improves the effectiveness of bioactives, broadens their application range and ensures optimal dosage, thereby improving cost-effectiveness of the product. Reactive or sensitive material, such as polynucleotides and polypeptides, can be turned into stable ingredients through encapsulation or entrapment by nanocarrier systems. It is also possible to prepare multireservoir nanocarriers in which two or more material are segregated in different compartments of the same capsule, minimising their contact and undesired interactions while releasing them at the target site simultaneously.

Because of their aforementioned properties, nanocarriers are a leading technology employed in nanotherapy to increase life standards of humans and animals. Nanocarrier systems vary in terms of ingredients, rigidity, stability, release properties and ability to incorporate materials with different solubilities. There is no universal nanocarrier that can be employed for all applications, although some are more versatile than others. The choice of which carrier system to use depends on the characteristics (size, solubility, charge, etc.) of the bioactive agent to be incorporated, safety and efficiency of the bioactive-carrier complex, as well as the intended application and route of administration of the complex to the body. With respect to manufacturing such complexes, possibility of mass production with minimum consumption of material, solvents, equipment and time should seriously be considered.

One of the challenges in nanotherapy is to reduce or completely eliminate side-effects. If bioactive agents act solely on their chosen target to produce the desired effect without causing unwanted effects on other systems, their usefulness will be enhanced significantly. In cancer therapy for example, in most cases, the drugs have no way of discriminating between normal dividing cells and the neoplastic cell line, making it difficult to completely eradicate cancer without at the same time doing irreparable damage to healthy cells and tissues. It is clear that nanotherapy would gain enormously if bioactives could be associated with target-specific nanocarrier systems. Targeted bioactive delivery can be defined as the ability of a given bioactive to concentrate in the required organ or tissue selectively and quantitatively, independent of the site and method of its administration [1]. Targeted delivery is under extensive research and development in order to increase bioactive concentration in the required sites with no harmful effects on normal healthy tissues, hence reducing the cost of therapeutic strategy [2, 3]. Different bioactive carrier and delivery technologies, along with their preparation, material incorporation and strategies for their targeting, are explained in this chapter.

2.   BIOACTIVE CARRIER SYSTEMS

A carrier system can be defined as the technology necessary for optimising the therapeutic efficiency of a bioactive agent. Ideally a carrier system should provide a nonreactive matrix to physically protect its load from destructive factors (e.g. oxygen, light, enzymes, etc.), physically separate it from other reactive components and also deliver it to the target site inside or outside the body. Carrier technologies can be broadly categorised as polymer, lipid and surfactant-based systems or a combination of these. Typical examples for carrier systems are microspheres [4, 5], nanospheres [6, 7], liposomes [8], nanoliposomes [9], vesicular phospholipid gels (VPG) [10], archaeosomes [11], and niosomes [12]. Examples of combined carrier systems are micelles [13] and nanoparticles [14] that can be prepared from polymer-lipid conjugates and polymer-surfactant mixtures respectively. In terms of size one can divide the carrier systems to micrometric (one to several micrometers) and sub-micrometric particles (nanocarriers), whose dimensions should be maintained from the point of preparation till the point of application. Some of these and other bioactive carrier technologies are explained in the chapters of this book.

In the design and formulation of carrier systems the key parameters are preparation and entrapment method, size, stability and degradation characteristics of the bioactive-carrier complex as well as release kinetics of the entrapped material. Generally biodegradable and bioabsorbable carriers are preferred so that they would degrade inside the body by hydrolysis or by enzymatic reactions, without producing any toxic product. More importantly, no preparation method will be useful unless it can be adapted to industrial scales for mass production of bioactive-carrier complexes with clearly defined reproducible properties. There are numerous lab-scale and a few large-scale techniques for the preparation of the carrier systems and entrapment of bioactives to these carriers. Some examples for these techniques and the resultant carriers are given in Table 1-1. Nevertheless, most of these methods are not suitable for entrapment of sensitive substances because of their exposure either to mechanical stresses (e.g. high-shear homogenisation, sonication, high pressures), potentially harmful chemicals (e.g. volatile organic solvents, detergents), or low/high values of pH during the preparation.

Table 1-1. Preparation techniques of some carrier systems.

Preparation Method	Carrier system	References
thin film hydration method	niosomes/liposomes	[9, 12, 15]
sonication	nanoliposomes/archaeosomes	[9]
heating method	nanoliposomes/liposomes/archaeosomes/VPG*	[9, 16-18]
reverse-phase evaporation	liposomes/archaeosomes	[9, 18]
ether-injection technique	liposomes	[9, 18]
high-pressure homogenisation	VPG*	[19]
precipitation polymerisation	nanoparticles	[20]
emulsification-diffusion method	nanoparticles	[21]
emulsion-solvent evaporation technique	microspheres	[22]

* vesicular phospholipid gels

Many of these preparation techniques involve the application of harmful volatile solvents (e.g. chloroform, ether, methanol, acetone) to dissolve or solubilise the ingredients. These solvents not only affect the chemical structure of the entrapped substance but also may remain in the final formulation and contribute to toxicity and influence the stability of the carrier systems. In general, residual solvents are known as organic volatile impurities and have no therapeutic benefits, but may be hazardous to human and animal health as well as the environment [23]. In addition to the above mentioned disadvantages, application of volatile organic solvents necessitates performance of two further steps in the preparation of the carrier systems: *i*) removal of these solvents, and *ii*) assessment of the level of residual volatile organic solvents in the formulations. Removal of these solvents from final preparations, especially in industry, requires application of costly and laborious vacuum procedures. Hence, avoiding the utilisation of these solvents and harmful procedures will bring down the toxicity as well as the time and cost of preparation of the carrier systems. For instance certain carriers, including liposomes, nanoliposomes, VPG and archaeosomes, can be prepared by *heating method* without employing volatile solvents or detergents and were found to be completely non-toxic when tested on human cultured cells [9, 16-18].

The cost of preparing carrier-entrapped bioactive molecules can be further reduced by minimising the number of steps and time it takes to produce these formulations into products that are suitable for human or animal use. It is also desirable to have a method capable of producing a vast number of polymer, lipid and surfactant-based carrier complexes in a single step and in a single piece of equipment or vessel, without involving harmful substances or procedures. Indeed, in a recent simple method, that is an improved and further developed version of heating method, different carrier systems (i.e. microspheres, nanospheres, liposomes, VPG, archaeosomes, niosomes, micelles and nanoparticles) can be prepared using a single apparatus in the absence of potentially toxic solvents in less than an hour [24]. The method is economical and capable of manufacturing bioactive carriers with a superior monodispersity and storage stability using a simple protocol. Another important feature of the method is that it can be adapted from small to industrial scales. This method is obviously most suitable for production of carrier systems for different in vitro and in vivo applications and involves heating and stirring the carrier ingredients, in the presence of a polyol, at a temperature between 50 to 120°C (based on the properties of the ingredients and material to be entrapped). Incorporation of bioactives into the carriers can be achieved by several routes including: *i*) adding the bioactive to the reaction medium along with the carrier ingredients and polyol; *ii*) adding the bioactive to the reaction medium when temperature has dropped to a point not lower than the transition temperature of the ingredients; and *iii*) adding the bioactive to the carrier after it is prepared at ambient temperatures (e.g. incorporation of different DNA molecules to micro- and nanocarriers can be achieved by this route [16, 18, 24-27]).

Following preparation of carrier systems and incorporation of bioactive materials the next issue to be addressed is targeting the formulation to its site of action, which can be inside or outside the body. Examples of in vitro bioactive targeting include delivery of encapsulated antibiotics, to control bacterial growth in systems such as food or pharmaceutical reactors, and cheese-ripening enzymes preferentially in the cheese curd for accelerated ripenning [28, 29]. Much of the research, however, has been devoted to devise systems for *in vivo* bioactive targeting as explained in the following sections.

3.   BARRIERS TO BIOACTIVE TARGETING

A logical design of optimal delivery systems entails considerations of various limiting steps in the *in vivo* and *in vitro* trafficking of bioactive agents. In the body, random and uncontrolled bioactive action on non-target tissues and failure of drugs to reach diseased cells often lead to devastating side-effects [30]. One of the main challenges in nanotherapy is to develop more selective, site-specific therapeutic agents for controlled and lasting treatment of diseases. Much emphasis is now being laid on the research and development of carrier systems that facilitate the delivery of the non-selective compounds to the diseased cells and organs preferentially, thereby imparting them with an exquisite target specificity and enhanced efficacy. Main impediments to bioactive targeting include physiological barriers as well as challenges to identify and validate the molecular targets and to devise appropriate techniques of conjugating targeting ligands to the nanocarriers [31]. The challenge in bioactive targeting is not only the targeting of bioactive to a specific site but also retaining it for optimum duration to elicit the desired action.

In pulmonary bioactive delivery the main points to consider for targeting the bioactive to a specific part of the lungs include particle size and density, the mucociliary escalator that clears the particles from the upper airways, clearance from the deep lung (alveoli) mainly by macrophage phagocytosis, clearance via the lymphatic system and uptake by alveolar type II epithelial cells.

For an intravenously administered nanosystem, the first and foremost barrier is that of the vascular endothelium and the basement membrane [32]. In addition, plasma proteins affect the biodistribution of carrier systems introduced into the blood. The *in vivo* biodistribution and opsonization of carriers in blood circulation is governed by their stability, size and surface characteristics. For a carrier to remain in blood circulation for a long time, the major problem is to avoid its opsonization and subsequent uptake by the phagocytic cells. The passage of bioactives and carrier systems across the endothelium depends on the molecular weight and size of the system, respectively. The tight endothelial cells in brain constitute the blood brain barrier (BBB) that restricts the entry of most molecules and their carriers. However, vascular endothelium is not uniform throughout. The altered endothelia in tumors allow an enhanced permeability to macromolecules and the particulate carrier systems. Another barrier is that of the extracellular matrix, which should be crossed to access the target cells in a tissue. If the whole tissue constitutes a target then the uniform distribution of bioactive throughout the tissue needs to be addressed [31, 32]. Targeting nanosystems to specific receptors or antigens on the cell surface provides the driving force for diffusion of the system to the specific cells.

A major parameter to consider in the formulation of efficient therapeutics is the solubility of bioactive compounds. Following administration of a bioactive to a human or animal, it must travel through multiple aqueous compartments, partitioned by lipid membranes, to reach its site of action in a particular organ or tissue. Furthermore, most of the bioactive compounds have their action sites in sub-cellular compartments such as cytosol or nucleus. However, the lipophilic nature of biological membranes restricts direct intracellular delivery of such compounds. Although there are several critical steps involved in transporting bioactives to the vicinity of the target cells, intracellular targeting is considered one of the biggest rate-limiting steps for bioactive delivery. For the bioactive agent to be transported while retaining its activity, it must exhibit reasonable solubility. However, as its aqueous solubility increases, its partition into biomembranes and its rate of diffusion across cellular membranes become proportionally difficult. Thus, the safe and efficient delivery of biologically active compounds to the eukaryotic cytosol has become one of the greatest challenges of the bioactive targeting.

The barriers do not end here, a number of endocytic pathways have been described for the cellular entry of nanocarriers. Bioactives/nanocarrier systems need to diffuse through the viscous cytosol to access particular cytoplasmic targets. Nuclear membrane poses another formidable barrier for bioactives such as oligonucleotides, plasmid DNA and low molecular weight drugs whose site of action is located in the cell nucleus. Although a number of cellular and molecular targets are emerging, the real problem lies with the poor accessibility of bioactives/nanocarriers to the target tissue. The presence of such barriers leads to a poor *in vitro/in vivo* correlation when the targeted delivery systems are tested in receptor bearing cells in vitro and fail *in vivo* [33]. Therefore, to exploit the potential of new targets in nanotherapy, one would need to develop targeted systems that can successfully overcome the physiological barriers and deliver the agent to its site of action at therapeutically relevant concentrations for a time sufficient to allow therapeutic action. Conjugation of targeting ligands to bioactives or carriers is the most applied method of directing them to their target sites. To this end, various techniques have been devised, including covalent and non-covalent conjugation [34]. The emphasis is that the ligand must be attached stably and accessibly to the drug carrier, so that the ligand is presented in its right orientation for binding to the target receptors. For example, the monoclonal antibodies must bind to the bioactive/carrier with their Fc part, so that their antigen-binding site (Fab) is free to interact with the antigenic targets on cells [35]. The coupling reactions must not affect the biological activity of ligand and should not adversely affect the structure of nanocarrier systems. Moreover, such coupling reactions must be optimized so that binding of ligands takes place in a homogeneous manner on the surface of the nano-carrier systems.

4.   BIOACTIVE TARGETING STRATEGIES

For a bioactive to exert its desired effect, it needs to be in physical contact with its physiological target, such as a receptor. Site-selective bioactive delivery ensures that such interactions take place only in the desired anatomical location of the body. Apparently, bioactives that already have an inherently high specificity for reaching and interacting with their targets (e.g. therapeutic antibodies) do not need to be considered for targeting. Mathematical analyses of bioactive targeting kinetics are provided in several publications [36-39], which also specify the properties needed for the site-specific delivery approach to work.

A possible approach to facilitate material transport into cells is the application of certain polypeptides such as bacterial pore-forming proteins [40], cell-penetrating peptides [41] or those capable of interacting with a particular type of tissue for transporting the bioactive to that tissue alone [42]. Examples of the later include hormones, antibodies and ricin A or B chain–transferrin conjugates [30]. A hallmark of these approaches is that a non-selective bioactive can be localized into a particular part of the body by the tissue specificity of the carrier molecule imparting the bioactive with target selectivity. Another strategy is the entrapment of bioactives in colloidal systems such as lipidic vesicles (e.g. liposomes, nanoliposomes, archaeosomes, VPG, etc.) [43]. Based on their solubilities, bioactives can be accommodated in the aqueous compartment(s) or lipophilic phases of these vesicles. Attachment of ligands such as sugars, hormones and antibodies endows these vesicular carriers with target specificity. A combination of both of the above-mentioned strategies – i.e. a polypeptide (e.g. TAT peptide) and a colloidal system (e.g. micelle, liposome or nanoparticle) – can also be employed in intracellular bioactive delivery [41].

Recently, acid- and salt-triggered multifunctional poly(propylene imine) dendrimers have been offered as a targetable delivery system [44]. The formulation contained poly(ethylene glycol) chains for stability and protection, as well as guanidium groups for targeting. The release of drugs is achieved through a change in pH. Another polymeric system claiming site-specific distribution is a water-soluble poly(vinylpyrrolidone-co-dimethyl maleic anhydride) [45] designed for bioactive delivery to kidneys. These carrier technologies deliver their loads to their targets mainly via passive or active targeting mechanisms as explained below.

4.1   Passive targeting

In passive targeting the bioactive-carrier complex reaches its destination by following the physio-anatomical conditions of the body. Site-specific delivery is achieved based on the physicochemical properties of bioactive-carrier complexes and does not require utilisation of any targeting strategy (see the next section on active targeting). The clearance kinetics and in vivo biodistribution of carrier systems depend on the physicochemical factors like size, charge and hydrophobicity and can be manipulated to enable passive targeting [46]. Generally when carriers are injected intravenously they are taken up by the reticulo-endothelial system (RES), mainly fixed macrophages in the liver and spleen after opsonization by proteins present in the blood stream [47]. Particles smaller than 100 nm can pass through the fenestrations in the liver endothelium and the sieve plates of sinusoids to localize in the spleen and bone marrow [31].

The natural tendency of carriers to localize in the RES presents an excellent opportunity for passive targeting of bioactives to the macrophages of the liver and the spleen. It has been utilized for targeting antibiotics and antiviral agents for intracellular infections and is very useful in the targeting of various diseases associated with the RES such as leishmaniasis and listeria. In these diseases the macrophages of the infected individual have a pathological role, therefore, if the macrophages are treated then the disease will be as well [48]. Once the carrier is engulfed by the macrophage, it will then be degraded and the drug will be released within the macrophage. Thus the drug will be delivered directly to the target area. In this scenario the carrier is allowed to follow its natural course yet at the same time it is able to deliver the drug passively to its target without any interference. Many investigators have exploited passive localization of parenterally administered colloidal carriers to infected macrophages of the RES for the delivery of antimicrobials [49]. It has been shown that the binding of anti-leishmanial drugs to various kinds of nanoparticles increase their therapeutic index (by 5 times) due to passive targeting of the RES [50].

The increased vascular permeability coupled with the impaired lymphatic drainage in tumors provides an opportunity for cancer targeting [51, 52]. For passive tumor accumulation the targeted system should have long blood circulation time, should not lose therapeutic activity while in circulation and the drug should stay with the carrier until accumulation of drug at the target is achieved. Other factors that may influence tumor targeting include the degree of tumor vascularization/angiogenesis and the size of the carrier system. Liposomes, polymeric nanoparticles, micellar systems as well as polymeric-drug conjugates have been successfully used to target drugs to tumor tissues in a passive manner [31]. Liposomes with diameters in the range of approximately 60-150nm, for example, can selectively extravasate in solid tumors [53, 54] thus exhibiting target specificity with negligible adverse effects to normal tissues.

Particle-size based passive targeting can also be applied in the formulation of inhalation products for bioactive delivery to the lungs. Deposition dose and site of inhaled particles within the lung depend largely on particle characteristics including size and shape. It is known that spherical material can be inhaled when its aerodynamic diameter is less than 10 micron. Generally, the smaller the particulates the deeper they can travel into the lung. Studies have shown that total lung deposition varies widely, depending on particle size and density as well as breathing pattern [55-58]. Particles smaller than 2.5 micron will even reach the alveoli. Ultrafine particles (nanoparticles with an aerodynamic diameter of <100 nm) are deposited mainly in the alveolar region [59].

Passive targeting can also be exploited for renal bioactive delivery. Poly(VP-co-DMMAn)-modified superoxide dismutase has been claimed to accumulate in the kidneys after intravenous administration and accelerate recovery from acute renal failure in a mouse model more effectively than superoxide dismutase alone [45]. This example utilizes the natural route by which an inert macromolecule is eliminated from the body via the kidneys.

4.2   Active targeting

Compared to passive targeting, active targeting offers more options for site-specific delivery of bioactive agents. Active targeting is achieved by engineering carriers sensitive to different stimuli (e.g. pH, temperature, light, etc.) or conjugating the bioactive/carrier system to one or more targeting ligands such as tissue or cell-specific molecules. Direct coupling of bioactives to targeting ligand restricts the coupling capacity to a few bioactive molecules. In contrast, coupling of nanocarriers to ligands allows import of thousands of bioactive molecules by means of one receptor targeted ligand. Bioactive targeting to specific cells has been explored utilizing the presence of various receptors and antigens/proteins on the cell membranes and also by virtue of the lipid components of the cell membranes. The receptors and surface bound antigens may be expressed uniquely in diseased cells only or may exhibit differentially higher expression in diseased cells as compared to the normal cells. Active agents, such as ligands for the receptors and antibodies to the surface proteins have been used extensively to target specific cells [31, 60]. Upon ligation to targeted receptors, both nanocarriers and the engaged receptors are often internalized by a complex process known as receptor-mediated endocytosis/phagocytosis [61]. The plasma membrane receptors considered for active targeting of bioactive agents include tuftsin, folate, transferrin and low density lipoprotein (LDL) receptors.

Lipid components of cellular membranes are emerging as novel targets for antineoplastic drugs [62]. Interaction of synthetic phospholipid analogs with cellular membranes changes the lipid composition, membrane permeability and fluidity, thereby influencing signal transduction mechanisms and inducing apoptotic cell death. Two such phospholipid analogs, i.e. Miltefosine and Edelfosine, can selectively kill malignant cells and thus offer promising approaches in cancer chemotherapy [31]. They have also been shown to have antileishmanial effects [63].

Active targeting can also be achieved utilising external stimuli such as ultrasound, magnetic field, or light. Ultrasound focused on tumor tissue has been shown to trigger the release of drugs from polymeric micelles only at the tumor site [64]. A decrease in tumor volume was demonstrated by focusing ultrasound on the tumors following an intravenous injection of Doxorubicin loaded micelles. Magnetic nanoparticles [65] and photo-sensitive liposomes [66] also offer exciting new opportunities towards developing effective bioactive targeting systems. It is possible to produce, characterize and specifically tailor the functional properties of these carrier technologies for clinical applications. A potential benefit of using magnetic nanoparticles is the use of localized magnetic field gradients to attract the particles to a chosen site, and the possibility to hold them there until the therapy is complete. Photosensitive liposomes on the other hand can release their contents at the target site following destabilisation of the liposomal membrane upon being subjected to localised light source [67].

The atypical conditions of cellular microenvironment and other physiological characteristics, such as local temperature increase in inflamed organs, can also be exploited to obtain accelerated bioactive release at a specific target by engineering particulates that are sensitive to these micro-environmental conditions [68]. Towards this end, particles incorporating pH-sensitive [69] or thermo-responsive [70] components have been synthesized to preferentially disintegrate at acidic pH or increased temperature, respectively. The sensitivity of thermo-responsive nanocarriers to elevated temperatures in the body can be useful in the treatment of solid tumors. Intracellular delivery of bioactives can be achieved using pH-sensitive nanocarriers, including pH-sensitive nanoliposomes, which provide non-invasive methods of targeting [41, 69]. The pH-sensitive carriers destabilize endosomal membrane under the low pH inside the endosome/lysosome compartment and liberate the entrapped bioactive into the cytoplasm.

5.   CONCLUDING REMARKS

Nanocarrier engineering is an essential requirement for the protection, segregation and delivery of bioactive material towards achieving optimal nanotherapy. Critical to the development of optimized bioactive delivery systems is the design of target-specific nanocarriers. For optimal therapeutic effect, an administered bioactive molecule must safely reach not only its target cell but also the appropriate location within that cell. Nanocarriers offer opportunities to achieve site-specific delivery with improved passive and active targeting mechanisms and newly discovered disease-specific targets. However, challenges remain, primarily because of the complexity of the body and the layers of barriers that these systems need to overcome to reach their target. To exploit the potential of new targets in nanotherapy, one would need to develop targeted systems that can successfully overcome the physiological barriers and deliver the agent to its site of action at therapeutically relevant concentrations for a time sufficient to allow therapeutic action. As nanotechnology for biomedical applications evolves, the safety of nanodevices including nanocarriers should also be investigated throughly.

6.   REFERENCES

1. Torchilin, V.P. (2005) Fluorescence microscopy to follow the targeting of liposomes and micelles to cells and their intracellular fate. *Adv. Drug Deliv. Rev.* 57: 95-109.
2. Francis, G.E. & Delgado, C. (Eds.) (2000) *Drug Targeting. Strategies, Principles, and Applications.* Humana Press, Totowa, NJ.
3. Torchilin, V.P. (2000) Drug targeting. *Eur. J. Pharm. Sci.* 11 (Suppl. 2): S81-S91.
4. Shen, C., Rao, P.V., Batich, C.D., Moorhead, J. & Yan, J. (1994) Stochastic modeling of controlled release from poly-styrene-co-4-vinylpyridine microspheres. *J. Control. Rel.* 32: 139-146.
5. Li, L., Song, H. & Chen, X. (2006) Hollow carbon microspheres prepared from polystyrene microbeads. *Carbon*, 44: 596-599.
6. Ravi Kumar, M.N.V., Bakowsky, U. & Lehr, C.M. (2004) Preparation and characterization of cationic PLGA nanospheres as DNA carriers. *Biomaterials*, 25: 1771-1777.
7. Teixeira, M., Alonso, M.J., Pinto, M.M.M. & Barbosa, C.M. (2005) Development and characterization of PLGA nanospheres and nanocapsules containing xanthone and 3-methoxyxanthone. *Eur. J. Pharm. Biopharm.* 59: 491-500.
8. Koromila, G., Michanetzis, G.P.A., Missirlis, Y.F. & Antimisiaris, S.G. (2006) Heparin incorporating liposomes as a delivery system of heparin from PET-covered metallic stents: effect on haemocompatibility. *Biomaterials*, 27: 2525-2533.

9. Mozafari, M.R. & Mortazavi, S.M. (Eds.) (2005) *Nanoliposomes: From Fundamentals to Recent Developments.* Trafford Pub. Ltd, Oxford, UK.

10. Farkas, E., Schubert, R. & Zelko, R. (2004) Effect of beta-sitosterol on the characteristics of vesicular gels containing chlorhexidine. *Int. J. Pharm.* 278: 63-70.

11. Patel, G.B., Omri, A., Deschatelets, L. & Sprott, G.D. (2002) Safety of archaeosome adjuvants evaluated in a mouse model. *J. Liposome Res.* 12: 353-372.

12. Palozza, P., Muzzalupo, R., Trombino, S., Valdannini, A. & Picci, N. (2006) Solubilization and stabilization of β-carotene in niosomes: delivery to cultured cells. *Chem. Phys. Lipids*, 139: 32-42.

13. Mu, L., Elbayoumi, T.A. & Torchilin, V.P. (2005) Mixed micelles made of poly(ethylene glycol)–phosphatidylethanolamine conjugate and d-α-tocopheryl polyethylene glycol 1000 succinate as pharmaceutical nanocarriers for camptothecin. *Int. J. Pharm.* 306: 142-149.

14. Mu, L. & Seow, P.H. (2006) Application of TPGS in polymeric nanoparticulate drug delivery system. *Colloid Surf, B: Biointerfaces*, 47: 90-97.

15. Pillai, G.K. & Salim, M.L.D. (1999) Enhanced inhibition of platelet aggregation in-vitro by niosome-encapsulated indomethacin. *Int. J. Pharm.* 193: 123-127.

16. Mozafari, M.R., Reed, C.J., Rostron, C., Kocum, C. & Piskin, E. (2002) Construction of stable anionic liposome-plasmid particles using the heating method: a preliminary investigation. *Cell. Mol. Biol. Lett.* 7: 923-927.

17. Mozafari, M.R., Reed, C.J. & Rostron, C. (2002) Development of non-toxic liposomal formulations for gene and drug delivery to the lung. *Technol Health Care*, 10 (3&4): 342-344.

18. Mozafari, M.R. (2005) Liposomes: an overview of manufacturing techniques. *Cell. Mol. Biol. Lett.* 10: 711-719.

19. Tardi, C., Drechsler, M., Bauer, K.H. & Brandl, M. (2001) Steam sterilisation of vesicular phospholipid gels. *Int. J. Pharm.* 217: 161-172.

20. Petrov, P., Rangelov, S., Novakov, C., Brown, W., Berlinova, I. & Tsvetanov, C.B. (2002) Core-corona nanoparticles formed by high molecular weight poly(ethylene oxide)-b-poly(alkylglycidyl ether) diblock copolymers. *Polymer*, 43: 6641-6651.

21. Quintanar-Guerrero, D., Fessi, H., Allemann, E. & Doelker, E. (1996) Influence of stabilizing agents and preparative variables on the formation of poly(D,L-lactic acid) nanoparticles by an emulsification-diffusion technique. *Int. J. Pharm.* 143: 133-141.

22. Huang, J., Liu, H., Gu, W., Yan, Z., Xu, Z., Yang, Y., Zhu, X. & Li, Y. (2006) A delivery strategy for rotenone microspheres in an animal model of Parkinson's disease. *Biomaterials*, 27: 937-946.

23. Dwivedi, A.M. (2002) Residual solvent analysis in pharmaceuticals. *Pharm. Technol. Europe*, 14: 26-28.

24. Mozafari, M.R. (2005) Method and apparatus for producing carrier complexes. *UK Patent No. GB 0404993.8, Int. Appl. No. PCT/GB05/000825 (03/03/2005).*

25. Zareie, M.H., Mozafari, M.R., Hasirci, V. & Piskin, E. (1997) Scanning tunnelling microscopy investigation of liposome-DNA-Ca^{2+} complexes. *J. Liposome Res.* 7(4): 491-502.

26. Mozafari, M.R. & Hasirci, V. (1998) Mechanism of calcium ion induced multilamellar vesicle-DNA interaction. *J. Microencapsul.* 15: 55-65.

27. Mozafari, M.R., Zareie, M.H., Piskin, E. & Hasirci, V. (1998) Formation of supramolecular structures by negatively charged liposomes in the presence of nucleic acids and divalent cations. *Drug Deliv.* 5: 135-141.

28. Kheadr, E.E., Vuillemard, J.C. & El Deeb, S.A. (2000) Accelerated Cheddar cheese ripening with encapsulated proteinases. *Int. J. Food Sci. Technol.* 35: 483-495.

29. Gouin, S. (2004) Micro-encapsulation: industrial appraisal of existing technologies and trends. *Trends Food Sci. Tech.* 15: 330-347.

30. Surolia, N. (2000) Receptor-mediated targeting of toxins to intraerythrocytic parasite Plasmodium falciparum. *Adv. Drug Deliv. Rev.* 41: 163-170.

31. Vasir, J.K., Reddy, M.K. & Labhasetwar, V.D. (2005) Nanosystems in drug targeting: opportunities and challenges. *Current Nanoscience*, 1: 47-64.

32. Garnett, M.C. (2001) Targeted drug conjugates: principles and progress. *Adv. Drug Deliv. Rev.* 53: 171-216.

33. Illum, L., Jones, P.D., Baldwin, R.W. & Davis, S.S. (1984) Tissue distribution of poly(hexyl 2-cyanoacrylate) nanoparticles coated with monoclonal antibodies in mice bearing human tumor xenografts. *J. Pharmacol. Exp. Ther.* 230: 733-736.

34. Nobs, L., Buchegger, F., Gurny, R. & Allemann, E.J. (2004) Current methods for attaching targeting ligands to liposomes and nanoparticles. *J. Pharm. Sci.* 93: 1980-1992.

35. Mathiowitz, E. (Ed.) (1999) *Encyclopedia of controlled drug delivery*. Wiley, New York.

36. Stella, V.J. & Himmelstein, K.J. (1985) Prodrugs: a chemical approach to targeted drug delivery. In: *Directed Drug Delivery: A Multidisciplinary Approach.* Borchardt, R.T., Repta, A.J. & Stella, V.J. (Eds.), Humana Press, pp. 247-267.

37. Hunt, C.A., MacGregor, R.D. & Siegel, R.A. (1986) Engineering targeted *in vivo* drug delivery. I. the physiological and physicochemical principles governing opportunities and limitations. *Pharm. Res.* 3: 333-344.

38. Smits, J.F.M. & Thijssen, H.H.W. (1986) Spatial control of drug action: theoretical considerations on the pharmacokinetics of target-aimed drug delivery. In: *Rate-Controlled Drug Administration and Action.* Struyker-Boudier, H.A.J. (Ed.), CRC Press, pp. 83-104.

39. Boddy, A., Aarons, L. & Petrak, K. (1989) Efficiency of drug targeting: steady-state considerations using a three-compartment model. *Pharm. Res.* 6: 367-372.

40. Provoda, C.J. & Lee, K.D. (2000) Bacterial pore-forming hemolysins and their use in the cytosolic delivery of macromolecules. *Adv. Drug Deliv. Rev.* 41: 209-221.

41. Gupta, B., Levchenko, T.S. & Torchilin, V.P. (2005) Intracellular delivery of large molecules and small particles by cell-penetrating proteins and peptides. *Adv. Drug Deliv. Rev.* 57: 637-651.

42. Pastan, I., Chaudhary, V.K. & Fitz Gerald, D.J. (1992) Recombinant toxins as novel therapeutic agents. *Annu. Rev. Biochem.* 1: 331-354.

43. Mozafari, M.R., Baran, E.T., Yurdugul, S. & Omri, A. (2005) Liposome-based carrier systems. In: *Nanoliposomes: From Fundamentals to Recent Developments*. Mozafari, M.R. & Mortazavi, S.M. (Eds.), Trafford Pub. Ltd, Oxford, UK, pp. 79-87.

44. Paleos, C.M., Tsiourvas, D., Sideratou, Z. & Tziveleka, L. (2004) Acid- and salt-triggered multifunctional poly(propylene imine) dendrimer as a prospective drug delivery system. *Biomacromolecules*, 5: 524-529.

45. Kamada, H., Tsutsumi, Y., Sato-Kamada, K., Yamamoto, Y., Yoshioka, Y., Okamoto, T., Nakagawa, S., Nagata, S. & Mayumi, T. (2003) Synthesis of a poly(vinylpyrrolidone-co-dimethyl maleic anhydride) co-polymer and its application for renal drug targeting. *Nat. Biotechnol.* 21: 399-404.

46. Stolnik, S., Illum, L. & Davis, S.S. (1995) Long circulating microparticulate drug carriers. *Adv. Drug Deliv. Rev.* 16: 195-214.

47. Jin, Y., Tong, L., Ai, P., Li, M. & Hou, X. (2006) Self-assembled drug delivery systems: 1. properties and *in vitro/in vivo* behavior of acyclovir self-assembled nanoparticles (SAN). *Int. J. Pharm.* 309: 199-207.

48. Daemen, T., Hofstede, G., Ten Kate, M.T., Bakker-Woudenberg, I.A. & Scherphof, G.L. (1995) Liposomal doxorubicin-induced toxicity: depletion and impairment of phagocytic activity of liver macrophages. *Int. J. Cancer*, 61 (5): 716-721.

49. Bakker-Woudenberg, I.A.J.M. (1995) Delivery of antimicrobials to infected tissue macrophages. *Adv. Drug Deliv. Rev.* 17: 5-20.

50. Rodrigues, Jr, J.M., Fessi, H., Bories, C., Puisieux, F. & Devissaguet, J.P. (1995) Primaquine-loaded poly(lactide) nanoparticles: physicochemical study and acute tolerance in mice. *Int. J. Pharm.* 126: 253-260.

51. McDonald, D.M. & Baluk, P. (2002) Significance of blood vessel leakiness in cancer. *Cancer Res.* 62 (18): 5381-5385.

52. Maeda, H., Wu, J., Sawa, T., Matsumura, Y. & Hori, K. (2000) Tumor vascular permeability and the EPR effect in macromolecular therapeutics: a review. *J. Control. Rel.* 65: 271-284.

53. Yuan, F., Leunig, M., Huang, S.K., Berk, D.A., Papahadjopoulos, D. & Jain, R.K. (1994) Microvascular permeability and interstitial penetration of sterically stabilized (stealth) liposomes in a human tumor xenograft. *Cancer Res.* 54: 3352-3356.

54. Sapra, P. & Allen, T.M. (2003) Ligand-targeted liposomal anticancer drugs. *Prog. Lipid Res.* 42: 439-462.

55. Heyder, J., Armbruster, L., Gebhart, J., Grein, E. & Stahlhofen, W. (1975) Total deposition of aerosol particles in the human respiratory tract for nose and mouth breathing. *J. Aerosol Sci.* 6: 311-328.

56. Chan, T.L. & Lippmann, M. (1980) Experimental measurements and empirical modeling of the regional deposition of inhaled particles in humans. *Am. Ind. Hyg. Assoc. J.* 41: 399-408.

57. Heyder, J., Gebhart, J., Rudolf, G., Schiller, C.F. & Stahlhofen, W. (1986) Deposition of particles in the human respiratory tract in the size range 0.005–15 μm. *J. Aerosol Sci.* 17: 811-825.

58. Stahlhofen, W., Rudolf, G. & James, A.C. (1989) Intercomparison of experimental regional aerosol deposition data. *J. Aerosol Med.* 2: 285-307.

59. Hoet, P.H.M., Bruske-Hohlfeld, I. & Salata, O.V. (2004) Nanoparticles – known and unknown health risks. *J. Nanobiotech.* 2 (1): 12.

60. Petrak, K. (2005) Essential properties of drug-targeting delivery systems. *Drug Discov. Today*, 10 (23/24): 1667-1673.

61. Moghimi, S.M. & Rajabi-Siahboomi, A.R. (2000) Recent advances in cellular, sub-cellular and molecular targeting. *Adv. Drug Deliv. Rev.* 41 (2): 129-133.

62. Jendrossek, V. & Handrick, R. (2003) Membrane targeted anticancer drugs: potent inducers of apoptosis and putative radiosensitisers. *Curr. Med. Chem. Anti-Canc. Agents*, 3: 343-353.

63. Zufferey, R. & Ben Mamoun, C. (2002) Choline transport in Leishmania major promastigotes and its inhibition by choline and phosphocholine analogs. *Mol. Biochem. Parasit.* 125: 127-134.

64. Rapoport, N., Pitt, W.G., Sun, H. & Nelson, J.L. (2003) Drug delivery in polymeric micelles: from in vitro to in vivo. *J. Control. Rel.* 91: 85-95.

65. Alexiou, C., Jurgons, R., Schmid, R.J., Bergemann, C., Henke, J., Erhardt, W., Huenges, E. & Parak, F. (2003) Magnetic drug targeting–biodistribution of the magnetic carrier and the chemotherapeutic agent mitoxantrone after locoregional cancer treatment. *J. Drug Target*, 11: 139-149.

66. Cirli, O.O. & Hasirci, V. (2004) UV-induced drug release from photoactive REV sensitized by suprofen. *J. Control. Rel.* 96: 85-96.

67. Wan, Y., Angleson, J.K. & Kutateladze, A.G. (2002) Liposomes from novel photolabile phospholipids: light-induced unloading of small molecules as monitored by PFG NMR. *J. Am. Chem. Soc.* 124: 5610-5611.

68. Jones, M.C. & Leroux, J.C. (1999) Polymeric micelles – a new generation of colloidal drug carriers. *Eur. J. Pharm. Biopharm.* 48: 101-111.

69. Huth, U.S., Schubert, R. & Peschka-Suss, R. (2006) Investigating the uptake and intracellular fate of pH-sensitive liposomes by flow cytometry and spectral bio-imaging. *J. Control. Rel.* 110: 490-504.

70. Na, K., Lee, K.H., Lee, D.H. & Bae, Y.H. (2006) Biodegradable thermo-sensitive nanoparticles from poly(l-lactic acid)/poly(ethylene glycol) alternating multi-block copolymer for potential anti-cancer drug carrier. *Eur. J. Pharm. Sci.* 27: 115-122.

Chapter 2

ARCHAEOSOMES AS DRUG AND VACCINE NANODELIVERY SYSTEMS

Girishchandra B. Patel and Wangxue Chen
Institute for Biological Sciences, National Research Council of Canada, Ottawa, Ontario, Canada K1A 0R6

Abstract: Archaeosomes, liposomes made from the polar lipids extracted from the membranes of *Archaea*, have been extensively studied for potential applications in drug and vaccine delivery over the past decade only. Archaeosomes offer significant advantages as nanodelivery vesicles over conventional liposomes made from the ester lipids found in *Eukarya* and *Bacteria*. The regularly branched, usually fully saturated isopranoid chains of archaeal polar lipids are attached to the *sn*-2, 3 carbons of the glycerol backbone(s) via ether bonds. In addition to monopolar lipids, as commonly found in eukaryotic and bacterial polar lipids, membrane-spanning bipolar lipids are encountered in archaeal species. These distinct features of archaeal polar lipids confer relatively superior physico-chemical stability properties to archaeosomes, such as thermal stability, stability in serum, stability at extremes of pH range, resistance to oxidative stress, and against the action of phospholipases and bile salts. Extensive *in vitro* and *in vivo* murine model studies indicate that archaeosomes are safe. These properties are beneficial for nanodelivery vesicle applications in biotechnology. Additionally, archaeosomes have been shown to be highly efficacious as self-adjuvanting vaccine delivery vesicles that promote robust, antigen-specific, humoral and cell-mediated immune responses, including $CD8^+$ CTL responses, to encapsulated protein antigens. These immune responses are well sustained over time, and are subject to strong memory responses. In murine models, archaeosome-delivered vaccines have demonstrated strong, long-lasting protective immunity against intracellular pathogens, as well as prophylactic and therapeutic efficacies against the development of experimental cancers. The archaeosome nanodelivery system is now ready for progressing from the laboratory to evaluations for actual commercial applications.

Key words: archaeosome, liposome, nanodelivery, drugs, vaccines, adjuvant, immunity

17

M.R. Mozafari (ed.), Nanocarrier Technologies: Frontiers of Nanotherapy, 17–40.
© 2006 *Springer. Printed in the Netherlands.*

1.   INTRODUCTION

It is only since the early 1990s that a significant effort has been made in studying the properties of liposomes made from the total polar lipid (TPL) extracts (as well as from purified polar lipids) of archaeal species for applications in drug and vaccine delivery [1]. Prior work only involved liposomes made from certain purified archael polar lipids in the context of understanding biomembrane functions. Liposomes made from polar ether lipids extracted from *Archaea*, or from lipid structures that mimic the unique structures of archaeal polar lipids, have been referred to as archaeosomes [1]. In this review, the terms "conventional liposomes" or "liposomes", without any further qualifications, refer to vesicles made from non-archaeal lipids, including vesicles made from ester lipid (or their ether lipid analogues) compositions with or without inclusion of additional components such as cholesterol. This article presents an introduction to archaeal polar lipid structures, the stability properties of archaeosomes, their tissue distribution and safety profile in murine models, as relevant to applications as drug and vaccine delivery systems. The potential utility of archaeosomes for developing protective vaccines against intracellular pathogens and for prophylactic/therapeutic cancer vaccines will be highlighted, using published data on murine models.

2.   *ARCHAEA* AND ARCHAEAL POLAR LIPIDS

Many members of *Archaea*, a domain of prokaryotes [2, 3], thrive in relatively harsh environments as compared to members in the domains *Eukarya* and *Bacteria*. Archaeal species include obligate anaerobes such as the methanogens, thermoacidophiles growing optimally at 55 to 80°C at pH values less than 3, hyperthermophiles growing optimally at >80°C, and extreme halophiles which need 4M NaCl for optimal growth. Despite these growth requirements, archaeal species are fairly ubiquitous in nature (gastrointestinal tracts of animals, humans and insects; coal refuse piles, salt lakes, bottom sediments of lakes and oceans, Antarctic and Alaskan waters, etc.), and in man-made environments such as anaerobic digesters. None of the archaeal species have been found to be pathogenic, and the typical Gram-negative outer membrane structures containing lipopolysaccharide (LPS) and porin-like components are not found in archaeal species [4-8].

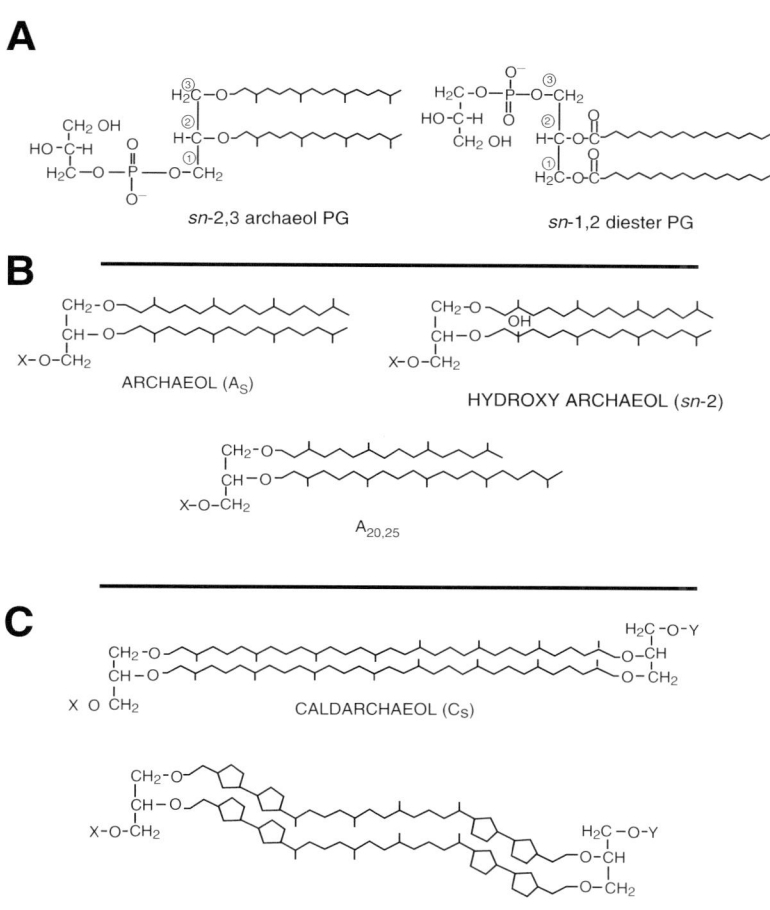

Figure 2-1. (A) Structural comparison between an archaeal diether phosphatidylglycerol (archaeol PG) lipid and a bacterial/eukaryotic diester phosphatidylglycerol (diester PG) lipid, (B) Core lipid structure of standard archaeol (A_s) and some variations thereof, (C) Core lipid structure of standard caldarchaeol (C_S) and variations thereof. In 1B and 1C, the nomenclature applies to the lipid cores where X and Y are protons. The polar lipids consist of the core lipid structures where X and Y represent the polar head groups.

One of the features that help distinguish *Archaea* from members of *Eukarya* and *Bacteria* is the unique structures of their polar membrane lipids [see reviews of [4-6, 9] for detailed structures]. The core structures (the non-polar hydrophobic portion) of archaeal polar lipids consist of regularly branched, 5-carbon repeating units that form the isopranoid chains (fully saturated with few exceptions) which are attached to the *sn*-2,3 glycerol carbons via ether bonds. In contrast to this, eukaryotic and bacterial phospholipids consist of un-branched fatty acyl chains (usually unsaturated) of variable lengths, attached to the *sn*-1, 2 glycerol carbons of the glycerol backbone via ester bonds (see comparison in Fig. 2-1A). Archaeal core lipid structures consist of the standard archaeol (2, 3-di-*O*-diphytanyl-*sn*-glycerol) consisting of 20 carbons per isopranoid chain (structure A$_s$ in Fig. 2-1B) and its modifications, and the standard caldarchaeol (2, 2', 3, 3'-tetra-*O*-dibiphytanyl-*sn*-diglycerol) consisting of 40 carbons per isopranoid chain (structure C$_s$ in Fig. 2-1C) and variations thereof (only some examples are shown). The polar lipids in archaeal species could consist exclusively of archaeol cores, or predominantly of caldarchaeol cores, or mixtures of both [5, 9]. The polar head groups (phospho, glyco, polyol, amino, hydroxyl) in archaeal lipids are similar to those found in eukaryotes and bacteria, however, phosphatidylcholine so common in ester lipids, is rarely observed in archaeal species. Archaeols are monopolar and caldarchaeols are bipolar. Archaeal TPL extracts have a net negative charge, due to the predominance of negatively charged polar head groups [5].

3.   ARCHAEOSOME PREPARATION

Archaeal biomass can be produced by growth in fermenter vessels, with certain precautions as appropriate for the species being grown [10, 11]. About 5% (w/w) of the biomass dry weight consists of the total lipid extract, of which approximately 90% consists of the acetone insoluble TPL fraction, and the remainder acetone soluble portion being the neutral lipids. The TPL can be extracted by solvent extraction methods [10, 11, 12], to obtain qualitative (as assayed by thin layer chromatography and by fast-atom bombardment mass spectrometry) and quantitative yields (actual yields of 1.3–3.6% dry wt of biomass, depending on the archaeal species; Patel, unpublished). The TPL can be stored for indefinite periods in chloroform, even in the presence of air, at room temperature or in the refrigerator, without chemical degradation.

The methods developed for production of unilamellar and multilamellar conventional liposomes [13-15], for the encapsulation/association of hydrophilic and hydrophobic compounds, are also applicable for making archaeosomes from the total lipids extracts (TPL plus neutral lipids), TPL extracts, or single polar lipids isolated in a biological pure form [11, 16-21]. It has been shown by thin layer chromatography that all the major identifiable lipids constituting the TPL extracted from specific archaeal species are incorporated into the respective archaeosome vesicles [22]. Examples of methods for making liposomes and archaeosomes include sonication, pressure extrusion protocols, freeze and thaw, detergent dialysis, reverse-phase evaporation, dehydration-rehydration and heating method. However, preparation of archaeosomes has lesser constraints as compared with making liposomes.

Preparation and storage of conventional liposomes from unsaturated ester lipids require measures such as use of inert atmospheres and/or addition of anti-oxidants during preparation and storage, to prevent lipid oxidation which can cause discoloration, leakage of encapsulated compounds, and production of toxic peroxides and aldehydes [14, 23]. Even the cholesterol, added at up to 33 mol% to help improve membrane stability of ester lipid liposomes [13], is prone to oxidation [23]. In contrast, the saturated isopranoid chains of archaeal polar lipids allow for the preparation and storage of archaeosomes in the presence of air/oxygen. Hydration of ester lipids is preferably done at temperatures above the phase transition (T_c) temperature of the predominant lipid component (to keep the lipids in a liquid crystalline state as opposed to the gel state), this temperature being higher than 20°C for many of the promising natural and synthetic ester lipids [13]. Archaeols and caldarchaeols are in the liquid crystalline state at 0°C and higher [24], and hence archaeosomes can be made at temperatures in the physiological range or lower, thus making it possible to encapsulate thermally labile compounds/antigens. In our laboratories, unilamellar archaeosome formulations with encapsulated low molecular weight markers [16] or ovalbumin, have been shown to be stable (respective to fusion, aggregation, leakage of encapsulated compound) for over 18 months of storage at 5°C, in the presence of air.

Unless indicated to the contrary, archaeosomes referred to in this review, were unilamellar vesicles of 60–300 nm average diameter, prepared from the TPL extracted from the indicated archaeal species. The convention used for describing archaeosome formulations in this article is as follows. "OVA/*M. smithii* archaeosomes" implies that ovalbumin (OVA) was encapsulated in unilamellar archaeosomes prepared from the TPL extracted from the archaeal species *Methanobrevibacter smithii*.

4.   SAFETY PROFILE

Earlier *in vitro* studies with unilamellar archaeosomes made from the purified major polar lipid (MPL) fraction of *T. acidophilum* suggested that they were safe, since they did not affect the growth or cellular functions of various cell lines and macrophages, or the synthesis of RNA, protein and DNA [25]. Archaeosomes made from the TPL of *M. smithii, Methano- spirillum hungatei* or *Methanosarcina mazei* also did not affect viability of J774A.1 macrophage cells [26].

Daily feeding of immunosuppressed NMRI mice with about 30 mg MPL lipid/kg body weight (blended with the regular diet) had no negative impact on the body weight or the longevity of mice [19]. Exhaustive evaluation upon daily oral gavage of BALB/c mice with much higher doses (55 to 550 mg/kg body weight per day, for 10 consecutive days) of empty archaeosomes prepared from 4 different archaeal species showed no dose related adverse effects on body temperature, body weight, % relative organ weights (heart, lungs, liver, spleen, kidneys), hepatic and renal function enzymes, or the parameters indicative of protein and carbohydrate metabolism [27]. These results were comparable to those obtained with feeding of ester liposomes made from a mixture of L-α-Dimyristoyl phosphatidylcholine (DMPC), L-α-Dimyristoyl phosphatidylglycerol (DMPG) and cholesterol [27]. These studies show that oral administration of archaeosomes is not harmful.

The short-term, repeated dose toxicity of six different, antigen-free archaeosome compositions, representative of TPL of two halophilic archaea, three methanogenic archaea, and a thermoacidophilic archaeon, was inves- tigated thoroughly, and compared with liposomes made from DMPC/ DMPG/CHOL [27]. BALB/c mice were administered the vesicles in PBS, at 0, 14, 70 or 140 mg lipid dry wt basis, per kg body weight, per day, for 5 consecutive days, and euthanized on the sixth day. Overall, there were no significant abnormal clinical signs or dose-dependent adverse effects except for the strong piloerection at the highest dose of the archaeosomes from TPL of the two halophiles (*H. salinarum* and *Natronobacterium magadii*) (Table 2-1), and a 23% decrease in the body weight and a drop in the average body temperature from 37.2°C to 31.6°C by the 3rd day (which slowly rose to 35°C on 6th day) in mice administered the highest dose of *H. salinarum* archaeosomes. Blood chemistry (blood urea nitrogen, glucose, alkaline phosphatase, alanine aminotransferase, aspartate aminotransferase) data did not indicate any dose-related toxicity either [27]. Some spleens from the high dose mice groups were enlarged and showed mild to moderate expansion of the red pulp [27]. Similar observations on spleens from mice intravenously administered conventional liposomes have also been reported previously [28, 29]. The intravenous dose range evaluated with archaeosomes [27] is similar to that assessed for liposomes, and the highest dose of 140 mg/kg/d is several folds greater than that anticipated as required for liposomal drug delivery applications [30]. Thus, except for the potential toxicity at the highest dose (140 mg/kg/d) of archaeosomes from the TPL of the halophilic archaeal species, archaeosomes seem to be relatively safe for i.v. applications.

Table 2-1. Piloerection and hyperactivity in mice intravenously administered ester liposomes or archaeosomes, at the indicated dose for 5 consecutive days.

Vesicle type[a]	Dose		
	(mg/kg/d)	Piloerection[b]	Hyperactivity[b]
Ester liposomes	14	N	N
	70	N	⊥ (3d)
	140	N	N
M. mazei	14	N	N
	70	N	N
	140	N	N
M. smithii	14	N	N
	70	N	N
	140	⊥ (2d)	⊥ (3d)
M. stadtmanae	14	N	⊥ (3d)
	70	N	⊥ (3d)
	140	N	+ (3d)
T. acidophilum	14	N	N
	70	N	N
	140	N	N
N. magadii	14	N	+ (4d)
	70	⊥ (2d)	+ (3d)
	140	++ (2d)	+ (3d)
H. salinarum	14	N	N
	70	⊥ (1d)	N
	140	++ (2d)	L (3d)

[a] Ester liposomes were made from DMPC/DMPG/CHOL; archaeosomes from the TPL of the indicated archaeal species
[b] N: no indication; ⊥: slight; +: moderate; ++: strong; L: lethargic; (d): beginning the day indicated
(Reproduced from Omri *et al.* [27], Copyright 2003, with permission from Taylor and Francis)

The safety of archaeosomes as relevant to vaccine applications was also evaluated in detail in mice [18]. Mice were subcutaneously immunized (0 and 21 d) with 11–20 μg OVA encapsulated in 1.25 mg archaeosomes (70 mg/kg body wt) prepared from the TPL of *M. smithii, T. acidophilum* or *H. salinarum.* Control groups received PBS, 11 μg OVA in PBS, or 1.25 mg empty archaeosomes in PBS. There were no injection site reactions (swelling, redness, bruising, granuloma, abscess formation), weight loss, or adverse clinical symptoms (behavior, piloerection, salivation, diarrhea, nasal discharge, alopecia) over the course of the 39-day study [18]. All other observations (blood chemistry and tissue histopathology) were comparable to those of the control groups. Additionally, compared with the control groups, there were no significant increases in the serum levels of creatine phosphokinase (CPK) in any of the archaeosome vaccinated groups at 1, 2, 22 and 39 days after immunization [18]. The archaeosome vaccine dose evaluated in this study [18] was about 125-fold greater than that needed for adjuvant efficacy, as discussed below.

Subcutaneous injection (1 mg/mouse/injection, total of 4 injections at 1-week intervals) of empty archaeosomes prepared from TPL of *M. smithii* or *T. acidophilum*, or empty DMPC/DMPG/CHOL liposomes, elicited little to no anti-lipid antibodies [18]. The biological significance of the low levels of anti-lipid antibodies (sera diluted only 1:50) detected in mice injected with *H. salinarum* archaeosomes is not clear [18]. *In vitro* hemolysis assays with mouse erythrocytes indicated that antigen-free *M. smithii, T. acidophilum* or *H. salinarum* archaeosomes caused no hemolysis at 0.3 to 1.25 mg/ml concentration [18].

The relatively good safety propfile of archaeosomes is not totally unexpected, since the C5 repeating units of the isopranoid chains in archaeal polar lipids are the same as those of the fat-soluble vitamins and Coenzyme Q_{10}. Additionally, ether-linked lipids present in plants, animals, and certain mammalian tissues [31], are part of the normal human diet. Further, liposomes made from ether lipid analogues of ester lipids (ether bonds at the *sn*-1,2 carbons of the glycerol backbone) have been shown to be metabolized and excreted in murine models [32, 33].

Although the potential metabolism of archaeosomes or archaeal polar lipids has not been specifically evaluated *in vivo*, cumulative evidence to-date showing vesicle susceptibility to lipases and bile salts [34], release of encapsulated compounds in macrophages [17,26], and delivery of encapsulated antigens for processing in the antigen processing cells, all suggest that they are being degraded. Initial studies, based on archaeosomes containing the non-metabolizable radioactive lipidic marker cholesteryl [1, 2-^{3}H] hexadecylether or ^{14}C-labeled *M. smithii* TPL, showed that bulk of the intravenously administered dose (30-60%) of *M. smithii* archaeosomes was excreted in the feces/urine by day 5 (G. B. Patel and S. Omar, unpublished). About 50-75% of similarly labeled archaeosomes injected into the mouse footpad (1 mg lipid per approximately 20 g mouse) were detected at the injection site 90 days post administration, and lesser quantities were detected in the liver, spleen and lymph nodes (G. B. Patel and S. Omar, unpublished). These preliminary data indicate that archaeosomes/archaeal lipids are indeed excreted. If they are degraded, the generation of toxic metabolic intermediates seems to be unlikely, based on the promising results discussed above on the general lack of toxicity of archaeosomes administered intravenously or orally [18, 27].

5. DRUG DELIVERY POTENTIAL

The structures of archaeal polar lipids suggest that archaeosomes prepared therefrom would have certain physico-chemical stability properties that would be advantageous for the delivery of drugs, imaging agents, and vaccines, as well as for *in vitro* applications. Similar to the bilayer membrane of liposomes made from ester phospholipids, archaeosomes prepared from archaeol lipids have a bilayer membrane. However, archaeosomes made from the bipolar caldarchaeol lipids form a monolayer that spans the vesicle membrane [35] and those made from a mixture of archaeol/caldarchaeol containing polar lipids consist of both the mono- and the bi-layer membrane structures. The rigidity provided by the membrane–spanning caldarchaeol lipids contribute to the stability of archaeosomes.

For oral delivery applications, the vesicles need to be stable not only in the low stomach pH, but they need to resist the disruptive effects of lipases and bile salts encountered in the gastrointestinal tract. At neutral pH (at 25°C), the stability of archaeosomes and conventional liposomes was comparable, but at pH 10, many archaeosome formulations demonstrated superior stability [16]. The stability of *T. acidophilum* multilamellar archaeosomes at pH 2.0 and 37°C (ca 60% intact vesicles after 24 h exposure) was comparable [34] to that of the most stable conventional liposomes made from a mixture of synthetic saturated ester lipid L-α-Distearoylphosphatidylcholine (DSPC) and cholesterol [36]. However, archaeosomes were considerably more resistant to the action of phospho-lipases (A_2, B, and pancreatic), than were conventional liposomes [16, 34]. Bile salts appeared to be similarly destructive to conventional liposomes [37, 38] and archaeosomes [34], but archaeosomes made from polar lipids high in caldarchaeol lipids seemed to be more stable [34]. Unlike multilamellar conventional liposomes where almost 100% of the encapsulated marker was released within 30 min of concurrent exposure to lipase and a bile salt [39], there was no such synergistic disruptive effect of pancreatic lipase and bile salts on archaeosomes [34]. After 1.5 h exposure of *T. acidophilum* multilamellar archaeosomes to the combined presence of sodium taurocholate (or simulated human bile) and phospholipase, about 50% of the encapsulated marker (Fig. 2-2) was still retained in the archaeosomes [34].

Although *in vivo* evaluation of the more stable caldarchaeol containing archaeosomes have not been conducted, studies with archaeosomes made exclusively from archaeol containing TPL from *Methanosarcina mazei* indicated that a large proportion (ca 28%) of the incorporated non-metabolizable radioactive lipidic marker was found in the stomach contents, stomach and the intestines at 6 h post oral gavage, and this dropped to about 10% and 1% after 24 and 48 h, respectively [40]. Very little of the marker (<1%) was detected in the blood or other organs at any of the time points beyond 0.25 h [40]. Incorporation of Coenzyme Q_{10} into orally administered archaeosomes resulted in an increased appearance of the marker in the blood [40].

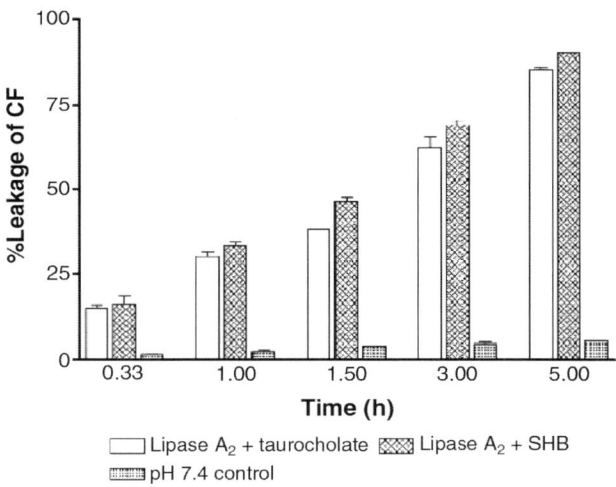

Figure 2-2. Percent leakage (mean ± SD) of encapsulated 5(6)-carboxyfluorescein (CF) from multilamellar vesicles prepared from *T. acidophilum* TPL, upon exposure (pH 7.4, 37°C) to a combination of phospholipase A_2 (5 U/ml) with either sodium taurocholate at 10 mg/ml or with simulated human bile (SHB). (Reproduced from Patel *et al.* [34], Copyright 2000, with permission from Elsevier).

Interaction of serum components with conventional liposomes [41] and archaeosomes made exclusively from archaeol lipids [16] can cause significant leakage of encapsulated compounds. However, archaeosomes made with TPL containing caldarchaeol lipids are much more stable, retaining about 80% of the encapsulated water-soluble marker after 5 h (37°C) exposure to serum [16, 42]. Upon intravenous (i.v.) administration, conventional liposomes are rapidly cleared from the blood stream, and are predominantly found in the liver and spleen shortly after administration [43,44]. Archaeosomes too are rapidly extravasated into the liver and spleen of mice upon i.v. administration [17, 19, 40]. Incorporation of polyethyleneglycol and Coenzyme Q_{10} into archaeosomes can alter the tissue distribution profiles of i.v. administered vesicles [40].

Archaeosomes were more thermo-stable at 4–65°C than were conventional liposomes, and even higher stability was associated with increasing proportion of caldarchaeol lipids in the TPL [16, 20, 45, 46]. The thermal stability of conventional liposomes too could be enhanced by the incorporation of caldarchaeol-containing polar lipids [42]. Archaeosomes made from many TPL compositions can be sterilized by autoclaving, without much loss of encapsulated, water-soluble low molecular weight marker [47]. This would help prepare sterile formulations, especially if the encapsulated compound is also amenable to autoclaving.

Although archaeosomes have not been directly evaluated in animal models for drug delivery applications, the superior stability properties of archaeosomes and *in vivo* studies to date, suggest possible advantages for such purposes. However, one needs to be cautious for applications in the delivery of protein/peptide drugs, considering the adjuvant potential of archaeosomes as discussed below.

6.   VACCINE DELIVERY APPLICATIONS

Archaeosomes prepared from different TPL compositions have the capacity to deliver the encapsulated antigen to antigen presenting cells (APCs) for MHC class I and class II presentation [48-50]. In contrast, conventional liposomes were unable to present OVA to the MHC class I pathway [50] and elicited only an antibody response that was generally inferior to that generated with archaeosomes [8, 51, 52]. It has been suggested that the archaeal lipid head group-APC receptor interaction, as well as the archaeal core lipid structures (branched, saturated isopranoid chains linked by ether bonds to the glycerol backbone) contribute to the adjuvanticity of archaeosome nanodelivery vehicles [49, 53]. The superior adjuvanticity of archaeosomes over conventional liposomes is perhaps largely due to their unique ability to act as immunomodulators, activating the innate immunity, and enhancing the recruitment and activation of professional APCs *in vivo*, rather than only due to the efficient delivery of the antigen to the APCs [8, 48, 51, 52, 54]. However, the robust memory responses could be speculated to be partly due to the longer antigen persistence that would be expected to be afforded by the depot effect of the stable archaeosomes [12, 16, 49, 51].

7.   HUMORAL IMMUNE RESPONSES

For generation of optimal antigen-specific responses to antigens encapsulated in conventional liposomes, the co-delivery of other known immunostimulants such as cholera toxin, lipid A or cytokines is required [55-57]. In contrast, archaeosomes are self-adjuvanting delivery vehicles which promote robust antigen-specific immunity in the absence of such requirements [51, 52]. For example, immunization of mice with bovine serum albumin (BSA) encapsulated in archaeosomes prepared from the TPL of various archaeal species induced BSA-specific serum antibody responses which were superior to that obtained with different conventional liposome formulations, and in some cases the titres were comparable to that generated with the potent, but highly toxic, Freund's complete adjuvant (Fig. 2-3) or with alum [51, 52, 8]. The efficacy of archaeosomes in the elicitation of humoral immune responses has also been proven using other model antigens such as cholera toxin B subunit, ovalbumin (OVA) and hen egg lysozyme (HEL), upon intraperitoneal (i.p.), intramuscular (i.m.) or subcutaneous (s.c.) immunization [48, 51, 52]. Archaeosomes prepared from the total lipid extracts (TPL plus the neutral lipids) or from the purified polar lipid fractions also promoted strong antibody responses [21, 58].

The efficacy of liposomes made from the ester lipids DMPC/DMPG to elicit humoral responses could also be enhanced by the incorporation of archaeal TPL in the vesicle composition [51]. Archaeosome vaccine-induced primary antibody responses were sustained for close to a year, and were amenable to strong memory recall upon antigen alone boost [51, 53, 59]. In contrast, the primary antibody titres induced by alum-adjuvanted vaccine had declined significantly by 200 days post-immunization, and the antibody memory boost was comparatively weaker [51]. The applicability of archaeosomes for delivery of multiple antigens (multivalent or combination vaccines) has also been demonstrated recently using OVA, HEL and BSA as the model antigens that were formulated into two types of archaeosome combination vaccines (Fig. 2-4). The anti-OVA, -HEL, and -BSA antibody responses obtained with the combination vaccine consisting of all three antigens co-encapsulated into one archaeosome formulation (CEC in Fig. 4) or with the admixed combination vaccine (AMC) obtained by admixing of three separately encapsulated antigen formulations, were similar to those obtained with the corresponding univalent vaccines (UVE) encapsulating either BSA, OVA or HEL [59]. The combination vaccines also demonstrated memory responses comparable to those obtained with the corresponding univalent vaccines [59].

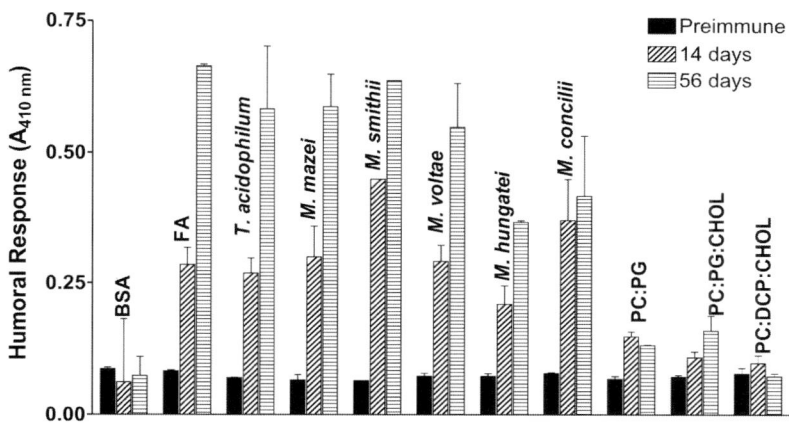

Figure 2-3. Comparison of murine humoral responses (IgG + IgM) to bovine serum albumin encapsulated in archaeosomes prepared from the TPL of indicated archaeal species, and in conventional liposomes made from the indicated mixtures of ester lipids (DMPC, DMPG, DCP) and cholesterol. Female BALB/c mice (6-8 weeks old) were injected intraperitoneally (25 µg BSA/injection) at days 0, 10 and 52, and sera assayed on the indicated days. Vesicle composition (mg lipid/boost): BSA, BSA in pH 7.1 PBS - no adjuvant or vesicles; FA, BSA emulsified in Freund's complete adjuvant in the first immunization and incomplete in the second, and BSA only (in PBS) in the third immunization; *T. acidophilum* (0.75 mg); *M. mazei* (0.64 mg); *M. smithii* (0.57 mg); *M. voltae* (1.83 mg); *M. hungatei* (1.09 mg); *M. concilii* (0.68 mg); PC:PG (DMPC:DMPG, 1.8:0.2 molar ratio, 2.11 mg); PC:PG:CHOL (DMPC:DMPG:cholesterol, 1.8:0.2:1.5 molar ratio, 0.12 mg); and PC:DCP:CHOL (DMPC:dicetyl-phosphate:cholesterol, 7:1:2 molar ratio, 0.87 mg). Anti-BSA antibody data are presented as mean ± standard deviation (A_{410} nm measuring IgG+IgM) in sera diluted 1:400. (Reproduced from Sprott *et al.* [52], Copyright 1997, with permission from Elsevier.)

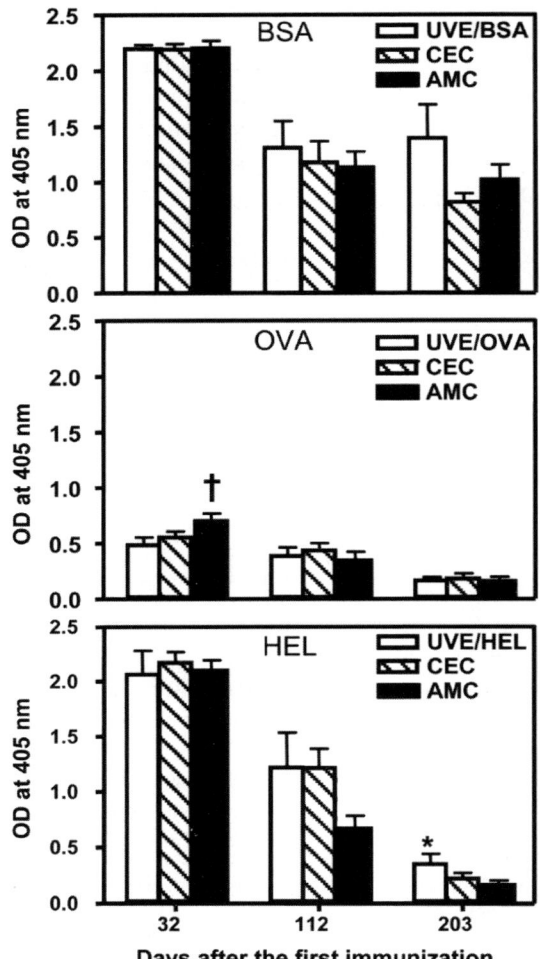

Figure 2-4. BSA- (top panel), OVA- (middle panel) and HEL- (bottom panel) specific antibodies in the sera of mice immunized with UVE/BSA (n = 5), UVE/OVA (n = 5), UVE/HEL (n = 5), CEC (n = 10) and AMC (n = 9) vaccines. Vaccines were given subcutaneously at day 0 and 21, and serum samples were collected at day 32, 112 and 203. The titres obtained using sera diluted 1:2000 are reported as OD_{405} ±SD. † $p < 0.05$ vs. UVE/OVA and CEC groups (32 d); * $p < 0.05$ vs. CEC and AMC groups (203 d). (Reproduced from Patel *et al*. [59], Copyright 2004, with permission from Francis and Taylor).

8.   CELL-MEDIATED IMMUNE RESPONSES

Archaeosomal vaccines generate predominantly IgG1 antibodies, significant amounts of IgG2a and IgG2b antibodies, but little to no IgM and IgG3 antibodies [17, 51], suggestive of the elicitation of both Th1 (IgG2a) and Th2 (IgG1) immune responses. The generation of robust cell-mediated immune responses to archaeosomally encapsulated proteins and peptides has been demonstrated in various mouse strains, including BALB/c (Th2 biased), C57BL/6 (Th1 biased), and C3H/HeJ (TLR4-defective) mice [51, 60]. Compared to the alum- or conventional liposome vaccine immunized mice, the splenocytes from archaeosome vaccinated mice demonstrated much stronger antigen-specific proliferation, significant *in vitro* production of Th1 (IFN-γ) and Th2 (IL-4) cytokines, as well as a high frequency of IFN-γ-secreting cells [8, 49, 51, 54, 60]. Therefore, archaeosomes stimulate both the humoral and cell-mediated arms of the immune system. Alum promotes exclusively a Th2 response, whereas conventional liposomes elicit a moderate antibody response and a much inferior cell-mediated response [8, 51, 52].

In contrast to conventional liposomes and alum adjuvanted vaccines which elicited little to no cytotoxic T lymphocyte (CTL) responses, archaeosome vaccines, made from the TPL (Fig. 2-5) or purified archaeal polar lipids, consisting of encapsulated proteins or peptides promoted strong and sustained antigen-specific CTL responses without augmenting the production of inflammatory cytokines IL-6 and IL-12 by DCs [12, 21, 50, 53, 60]. Even a single s.c. immunization with OVA encapsulated into various archaeosomes promoted good primary CTL responses at day 10 [12, 48, 50]. Archaeosome induced CTL responses were generally well sustained [50] and some formulations (e.g. *M. smithii* or *T. acidophilum* archaeosomes) demonstrated potent CTL memory responses even at close to a year subsequest to the primary immunization [12]. The CTL response elicited by archaeosomes was CD8[+] T cell-dependent and perforin-mediated [50]. Additionally, archaeosomes can elicit a potent CD4[+] T cell-independent CD8[+] T cell response [50]. An efficient maintenance of T-cell memory [51], and long-term CTL memory response generated by archaeosomes is correlated to strong Ag-specific up-regulation of CD44 on splenic CD8[+] T cells [50].

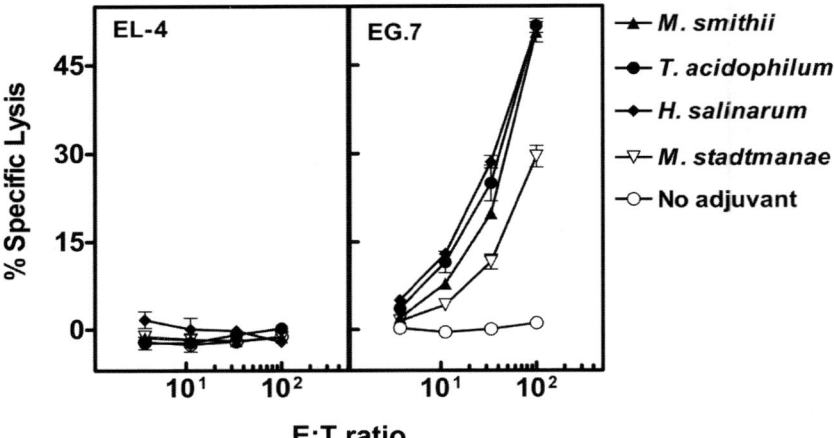

Figure 2-5. Induction of CTL responses by archaeosomes composed of divergent lipid compositions. C57BL/6 mice were immunized (s.c.) with 15 μg OVA either in PBS (no adjuvant) or entrapped in archaeosomes composed of TPL from *M. smithii*, *T. acidophilum*, *H. salinarum*, or *M. stadtmanae*. On day 14 the spleens were harvested, pooled (n = 3/group), and stimulated with irradiated EG.7 cells for 5 days, and then 4-h CTL activity on ^{51}Cr-labeled targets (EL-4 and EG.7) was assessed. Percent specific lysis ± SD of triplicate cultures is indicated at various E: T ratios. (Reproduced from Krishnan *et al.* [50], Copyright 2000, The American Association of Immunologists, Inc.).

9.   EFFICACY IN INTRACELLULAR INFECTION MODEL

The host elicitation of antigen-specific cell-mediated immunity, especially CD8^{+} T cell-mediated CTL responses, required for conferring protective immunity against intracellular pathogens such as *Plasmodium spp.*, *Leishmania spp.*, viruses, *Mycobacterium tuberculosis*, and *Listeria monocytogenes* [61], has typically only been achieved by vaccination with live, attenuated pathogens. However, due to the regulatory concerns with the use of live pathogens, there is a great interest in developing appropriately adjuvanted, defined, acellular (sub-unit) vaccines that are equally efficacious. The proof of principle on the applicability of archaeosomes for developing such vaccines was evaluated using a murine model of *L. monocytogenes* infection, since it is well established that long-term protective immunity against this pathogen is for the most part mediated by $CD8^+$ T cells, although $CD4^+$ T cells and, to a lesser extent, antibodies also contribute to the protective immunity [62].

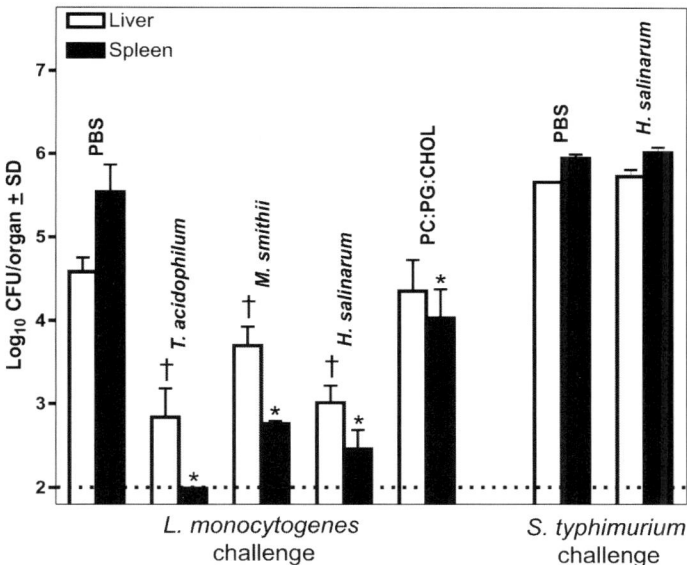

Figure 2-6. Burdens of *Listeria* and *Salmonella* in the livers and spleens of control mice and mice vaccinated one week earlier with lipopeptide antigen encapsulated in various vesicle types. BALB/c mice (n = 5/group) were vaccinated once, with 15 μg of the 20-mer lipopeptide (protective antigen for *Listeria*) encapsulated in 0.37 mg of *T. acidophilum*, 0.67 mg of *M. smithii*, or 0.60 mg of *H. salinarum* archaeosomes, or 0.54 mg of conventional liposomes (DMPC/DMPG/CHOL), or with only PBS (control group), as indicated. One week post-vaccination, mice were challenged intravenously with either the target pathogen *L. monocytogenes* (2.5 × 10^3 CFU per mouse) or the heterologous intracellular pathogen *Salmonella typhimurium* (3.1 × 10^3 CFU/mouse). Bacterial burdens in the livers and spleens (Log$_{10}$ ± SD, CFU/organ) were determined on day three of infection.
†, * significantly lower ($P < 0.05$) than that in the corresponding organs of control group, by Mann and Whitney rank sum test. ---- line indicates the lower detection limit in the bacterial enumeration. (Plotted from Table 4 of Conlan *et al.* [60]).

The peptide antigens used in these studies (bipalmitoylated 13-mer or 20-mer peptide) contained the highly immunodominant nonameric MHC class I restricted CD8$^+$ T-cell-specific listeriolysin epitope for BALB/c mice [63]. Mice vaccinated with the lipopeptide (s.c. at 0, 21 and 42 d, 12.5 µg of LLO-derived 20-mer lipopeptide) encapsulated in *M. smithii* archaeosomes were well protected upon i.v. challenge with *L. monocytogenes* (2.4 × 10^3 colony forming units/mouse) at 28 days post last immunization. The pathogen burdens were up to 38-fold lower in the livers and up to 2042-fold lower in the spleens, compared to the burdens in corresponding organs from the control (naïve, or mice immunized with lipopeptide alone or empty archaeosomes or with empty archaeosomes admixed with the lipopeptide) groups [60]. This protective immunity was long lasting, being still evident at up to 10 months (last assayed point) post-vaccination [60, 64]. Even a single s.c. immunization of mice with the lipopeptide encapsulated in either of three different archaeosomes elicited a robust protective immunity (ca two log$_{10}$ reduction of pathogen burden in livers; up to three log$_{10}$ reduction in spleens, with many spleens being free of the pathogen) as early as day 7 (Fig. 2-6). This immunity was superior to that attained with the ester lipid liposome formulation (DMPC/DMPG/CHOL), and was antigen-specific (no protection of archaeosome vaccinated mice upon challenge with the heterologous intracellular pathogen *Salmonella typhimurium*). Additionally, mice vaccinated with as little as 0.5 µg of the lipopeptide encapsulated in 10 µg archaeosomes developed protective immunity at 7 days post a single s.c. immunization [64]. Unlike conventional liposomes which required co-encapsulation of the immunostimulant Quil-A for eliciting anti-listeria immunity [65], archaeosome vaccines promoted rapid and long lasting protective immunity without the need for including other immunostimulants. These results demonstrate the great potential for archaeosome applications in developing protective, non-replicating, defined, acellular vaccines against intracellular pathogens.

10.   EFFICACY IN CANCER MODELS

Generation of antigen-specific CD8$^+$ CTL responses are critical for protection against the development of primary tumors and their metastasis, and for the elimination of established neoplastic cells [61]. Therefore, with the known profile of the immune responses elicited by archaeosomes, it was logical to determine their efficacy for prophylactic and therapeutic cancer vaccines. The prophylactic efficacy of archaeosome-adjuvanted vaccines in protecting against metastatic cancer was evaluated in a murine model of lung metastasis [54] using a H-2^b restricted OVA-transfected melanoma cell line

(B16OVA). Naïve mice and mice immunized (s.c., at 0 and 21 d) with OVA/*M. smithii* archaeosomes were challenged i.v. with 5×10^5 B16OVA cells on day 35, and the lungs from all groups were examined at 14 days post-challenge for the presence of tumor foci [54]. In contrast to the lungs of the OVA/archaeosome immunized mice which were virtually free of tumor foci, the lungs of naïve mice had 250–500 macroscopically visible black tumour foci [54]. Similar protective immunity with archaeosome vaccines was also obtained when a naturally occurring melanoma differentiation antigen TRP2 (tyrosinase related protein-2) was used as the immunogen [12]. These results suggested the utility of archaeosomes in prophylactic vaccine applications to elicit immune response to tumor antigens.

The evaluation in the primary (solid) tumor model was based on the use of OVA-transfected T cell lymphoma EG.7 cells [66]. Female C57BL/6 mice immunized with OVA encapsulated in *M. smithii* archaeosomes were tumor-free [54] even at 20 d post-challenge with EG.7 cells (Fig. 2-7). In contrast to this, mice in all the three control groups developed tumors of approximately 300 mm^2 size by 12-15 days post-challenge (Fig. 2-7), and had to be euthanized [54]. More than 50% of OVA/archaeosome-immunized mice were tumor-free at 60 days, while the rest had small tumors appearing at between 28 and 42 days [54]. Similar protection was obtained with vaccines comprising of *T. acidophilum* or *H. salinarum* archaeosomes [12]. It was also observed that OVA/*M. smithii* archaeosomes were able to induce IL-12-independent tumor protective CD8$^+$ CTL responses [54].

The potential of archaeosomes as an adjuvant for therapeutic cancer vaccines has also been examined using the OVA-expressing EG.7 murine tumor model. Mice were first injected in the mid-back region with approximately 6×10^6 EG.7 cells, followed by s.c. immunization with 15 μg OVA encapsulated in *M. smithii* archaeosomes at days 1 and 7 post-tumor injection [54]. The OVA/archaeosome immunized mice had substantially suppressed tumor growth (generally <100 mm^2 size) or even induced tumor regression, whereas the tumor growth progressed rapidly to over 200 mm^2 size by 12–15 days in the naïve mice or mice immunized with OVA only [54]. Similar therapeutic protection was obtained in mice immunized with OVA/*H. salinarum* or OVA/*T. acidophilum* archaeosomes [12]. These results clearly demonstrate the applicability of archaeosomes in therapeutic cancer vaccine development.

In the EG.7 tumor model therapeutic studies, it was observed that even immunization with empty (i.e. antigen-free) *M. smithii* or *T. acidophilum* archaeosomes at 0 and 7 to 10 day of challenge resulted in substantially diminished EG.7 tumor growth [12, 54]. However, a similar effect was not obtained with antigen-free conventional liposomes made from DMPC/ DMPG/CHOL [54] or *H. salinarum* archaeosomes [12]. Considering that antigen-free archaeosomes have no prophylactic effect on tumor growth [54], the therapeutic suppression of tumor progression by the two antigen-free archaeosome formulations suggested that some archaeosomes may facilitate the exertion of a short-term protective innate immunity [12, 54].

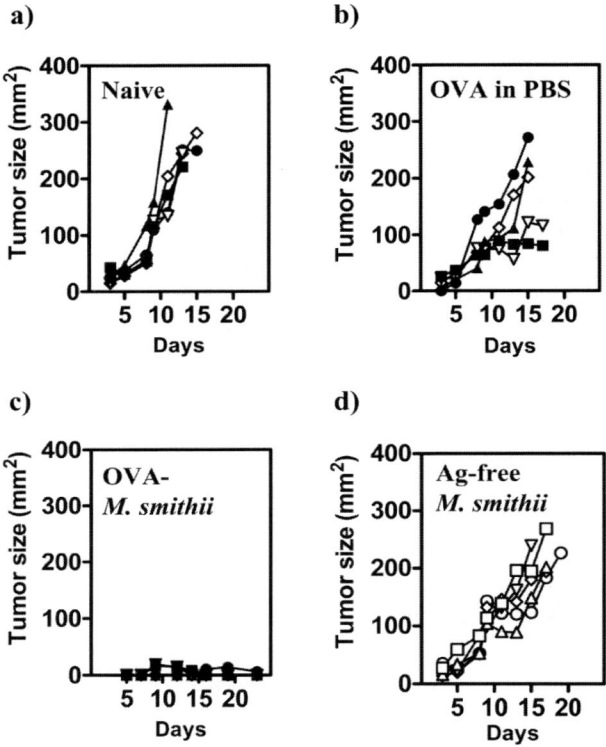

Figure 2-7. Prophylactic vaccination against solid tumors using OVA entrapped in archaeosomes. C57BL/6 mice were immunized on days 0 and 21 with 15 µg OVA entrapped in 227 µg of *M. smithii* archaeosomes (c). Controls included age-matched naïve mice (a), those injected with OVA in the absence of adjuvant (b), and those injected with 227 µg of *M. smithii* archaeosomes lacking OVA (d). On day 42, the mice were challenged s.c. in the mid-dorsal region with 7×10^6 EG.7 cells. Solid tumor progression was monitored using digital calipers. Each line represents tumor progression in individual mouse. Tumor size is represented as L $\times$ W (mm^2). Tumor sizes in the OVA/*M. smithii*-immunized group were significantly different ($P < 0.0001$) from naïve controls and antigen-free archaeosomes, as analyzed by ANOVA. These data are representative of six independent experiments. (Reproduced from Krishnan *et al.* [54], with permission from The American Association of Cancer Research, Inc.)

11.   SYNOPSIS

In just over a decade, it has been demonstrated that archaeosomes represent a safe nanodelivery system for the delivery of drugs and vaccines. The structural features of the archaeal polar lipids allow for the preparation of stable archaeosomal formulations in the absence of the normal precautions required (e.g. protection from air to avoid oxidation) for conventional ester liposome formulations. The knowledge about archaeosome applications in the vaccine field is much more advanced than for drug delivery. The advantages of archaeosomal vaccines are that they represent non-replicating, self-adjuvanting, antigen delivery systems which can promote potent, long lasting, antigen-specific humoral and cell-mediated immunities (including $CD8^+$ CTL), and elicit strong memory responses. The efficacy of archaeosome adjuvants has been validated in murine models of intracellular infection and in cancers. Studies to-date indicate that the immune profiles promoted by archaeosome adjuvanted protein and peptide vaccines approximate those currently achieved only with live, replicating vaccines. Archaeosome technology has advanced to the stage where it is now ready for evaluation for actual commercial applications.

Acknowledgements

We acknowledge the contributions of our colleagues in advancing the knowledge on archaeosomes. This is NRCC publication 42507.

12.   REFERENCES

1. Patel, G. B. and Sprott, G. D. (1999). *Crit Rev Biotechnol* **19,** 317-357.
2. Boone, D. R., Whitman, W. B., and Rouviere, P. (1993). In *Methanogenesis*, Ferry, J. G. (Ed.), Chapman & Hall, New York, pp. 35-80.
3. Madigan, M. T. and Marrs, B. L. (1997). *Sci Am* **276,** 82-87.
4. De Rosa, M. and Gambacorta, A. (1988). *Prog Lipid Res* **27,** 153-175.
5. Sprott, G. D. (1992). *J Bioenerg Biomembr* **24,** 555-566.
6. Kates, M. (1997). *Am Chem Soc Symp Series* **671,** 35-47.
7. Gophna, U., Charlebois, R. L., and Doolittle, W. F. (2004). *Trends Microbiol* **12,** 213-219.
8. Krishnan, L., Sad, S., Patel, G. B., and Sprott, G. D. (2001). *J Immunol* **166,** 1885-1893.
9. Kates, M. (1992). *Biochem Soc Symp* **58,** 51-72.

10. Sprott, G. D., Choquet, C. G., and Patel, G. B. (1995). In *Methanogens. Archaea-A Laboratory Manual*, Sowers, K. R. and Schreier, H. J. (Eds), Cold Spring Harbor Laboratory Press, New York, pp. 329-340.

11. Sprott, G. D., Patel, G. B., and Krishnan, L. (2003). *Methods Enzymol* **373,** 155-172.

12. Krishnan, L. and Sprott, G. D. (2003). *J Drug Target* **11,** 515-524.

13. Szoka, F., Jr. and Papahadjopoulos, D. (1980). *Annu Rev Biophys Bioeng* **9,** 467-508.

14. Betageri, G. V., Jenkins, S. A., and Parsons, D. L. (1993). *Liposome Drug Delivery Systems*, Technomic Publishing Co., Inc., Lancaster, PA.

15. Watwe, R. M. and Bellare, J. R. (1995). *Curr Sci* **68,** 715-724.

16. Choquet, C. G., Patel, G. B., Beveridge, T. J., and Sprott, G. D. (1994). *Appl Microbiol Biotechnol* **42,** 375-384.

17. Makabi-Panzu, B., Sprott, G. D., and Patel, G. B. (1998). *Vaccine* **16,** 1504-1510.

18. Patel, G. B., Omri, A., Deschatelets, L., and Sprott, G. D. (2002). *J Liposome Res* **12,** 353-372.

19. Freisleben, H. J., Bormann, J., Litzinger, D. C., Lehr, F., Rudolph, P., Schatton, W., and Huang, L. (1995). *J Liposome Res* **5,** 215-223.

20. Ring, K., Henkel, B., Valenteijn, A., and Gutermann, R. (1986). In *Liposomes as Drug Carriers*, Schmidt, K. H. (Ed.), Thieme, Stuttgart, pp. 101-123.

21. Sprott, G. D., Patel, G. B., Makabi-Panzu, B., and Tolson, D. L. (1997). *PCT International Publication No.* WO97/22333.

22. Choquet, C. G., Patel, G. B., Beveridge, T. J., and Sprott, G. D. (1992). *Appl Environ Microbiol* **58,** 2894-2900.

23. Ausborn, M., Nuhn, P., and Schreier, H. J. (1992). *Eur J Pharm Biopharm* **38,** 133-139.

24. Blocher, D., Six, L., Gutermann, R., Henkel, B., and Ring, K. (1985). *Biochim Biophys Acta* **818,** 333-342.

25. Freisleben, H. J., Neisser, C., Hartmann, M., Rudolph, P., Geck, P., Ring, K., and Muller, E. G. (1993). *J Liposome Res* **3,** 817-833.

26. Tolson, D. L., Latta, R. K., Patel, G. B., and Sprott, G. D. (1996). *J Liposome Res* **6,** 755-776.

27. Omri, A., Agnew, B. J., and Patel, G. B. (2003). *Int J Toxicol* **22,** 9-23.

28. Claassen, E., Westerhof, Y., Versluis, B., Kors, N., Schellekens, M., and van Rooijen, N. (1988). *Br J Exp Pathol* **69,** 865-875.

29. Allen, T. M., Murray, L., MacKeigan, S., and Shah, M. (1984). *J Pharmacol Exp Ther* **229,** 267-275.

30. Storm, G., Oussoren, C., Peeters, P. A. M., and Barenholz, Y. (1993). In *Liposome Technology. Vol. 3. Interaction of Liposomes with the Biological Milieu.*, Gregoriadis, G. (Ed.), CRC Press, Boca Raton, pp. 345-383.

31. Snyder, F. (1991). In *Biochemistry of Lipids, Lipoproteins and Membrances.*, Vance, D. E. and Vance, J. (Eds), Elsevier Science Publishers B.V., Amsterdam, pp. 241-267.

32. Stein, Y., Halperin, G., Leitersdorf, E., Dabach, Y., Hollander, G., and Stein, O. (1984). *Biochim Biophys Acta* **793,** 354-364.

33. Derksen, J. T., Baldeschwieler, J. D., and Scherphof, G. L. (1988). *Proc Natl Acad Sci U S A* **85,** 9768-9772.

34. Patel, G. B., Agnew, B. J., Deschatelets, L., Fleming, L. P., and Sprott, G. D. (2000). *Int J Pharm* **194,** 39-49.

35. Beveridge, T. J., Choquet, C. G., Patel, G. B., and Sprott, G. D. (1993). *J Bacteriol* **175,** 1191-1197.

36. Aramaki, Y., Tomizawa, H., Hara, T., Yachi, K., Kikuchi, H., and Tsuchiya, S. (1993). *Pharm Res* **10,** 1228-1231.

37. Richards, M. H. and Gardner, C. R. (1978). *Biochim Biophys Acta* **543,** 508-522.

38. Rowland, R. N. and Woodley, J. F. (1980). *Biochim Biophys Acta* **620,** 400-409.

39. Chiang, C.-M. and Weiner, N. (1987). *Int J Pharm* **37,** 75-85.

40. Omri, A., Makabi-Panzu, B., Agnew, B. J., Sprott, G. D., and Patel, G. B. (2000). *J Drug Target* **7,** 383-392.

41. Allen, T. M. and Cleland, L. G. (1980). *Biochim Biophys Acta* **597,** 418-426.

42. Sprott, G. D., Dicaire, C. J., Fleming, L. P., and Patel, G. B. (1996). *Cells and Materials* **6,** 143-155.

43. Senior, J. H. (1987). *Crit Rev Ther Drug Carrier Syst* **3,** 123-193.

44. Allen, T. M. (1994). *Adv Drug Del Rev* **13,** 285-309.

45. Patel, G. B., Agnew, B. J., Jarrell, H. C., and Sprott, G. D. (1999). *J Liposome Res* **9,** 229-245.

46. Elferink, M. G., de Wit, J. G., Driessen, A. J., and Konings, W. N. (1994). *Biochim Biophys Acta* **1193,** 247-254.

47. Choquet, C. G., Patel, G. B., and Sprott, G. D. (1996). *Can J Microbiol* **42,** 183-186.

48. Sprott, G. D., Sad, S., Fleming, L. P., Dicaire, C. J., Patel, G. B., and Krishnan, L. (2003). *Archaea* **1,** 151-164.

49. Gurnani, K., Kennedy, J., Sad, S., Sprott, G. D., and Krishnan, L. (2004). *J Immunol* **173,** 566-578.

50. Krishnan, L., Sad, S., Patel, G. B., and Sprott, G. D. (2000). *J Immunol* **165,** 5177-5185.

51. Krishnan, L., Dicaire, C. J., Patel, G. B., and Sprott, G. D. (2000). *Infect Immun* **68,** 54-63.

52. Sprott, G. D., Tolson, D. L., and Patel, G. B. (1997). *FEMS Microbiol Lett* **154,** 17-22.

53. Sprott, G. D., Dicaire, C. J., Gurnani, K., Deschatelets, L. A., and Krishnan, L. (2004). *Vaccine* **22,** 2154-2162.

54. Krishnan, L., Sad, S., Patel, G. B., and Sprott, G. D. (2003). *Cancer Res* **63,** 2526-2534.

55. Richards, R. L., Rao, M., Wassef, N. M., Glenn, G. M., Rothwell, S. W., and Alving, C. R. (1998). *Infect Immun* **66,** 2859-2865.

56. Harokopakis, E., Hajishengallis, G., and Michalek, S. M. (1998). *Infect Immun* **66,** 4299-4304.

57. Lachman, L. B., Ozpolat, B., and Rao, X. M. (1996). *Eur Cytokine Netw* **7,** 693-698.

58. Sprott, G. D., Brisson, J., Dicaire, C. J., Pelletier, A. K., Deschatelets, L. A., Krishnan, L., and Patel, G. B. (1999). *Biochim Biophys Acta* **1440,** 275-288.

59. Patel, G. B., Zhou, H., KuoLee, R., and Chen, W. (2004). *J Liposome Res* **14,** 191-202.

60. Conlan, J. W., Krishnan, L., Willick, G. E., Patel, G. B., and Sprott, G. D. (2001). *Vaccine* **19,** 3509-3517.
61. Pardoll, D. M. (1998). *Nat Med* **4,** 525-531.
62. Pamer, E. G. (2004). *Nat Rev Immunol* **4,** 812-823.
63. Vijh, S., and Pamer, E. G. (1997). *J Immunol* **158,** 3366-3371.
64. Sprott, G. D., Krishnan, L., Conlan, J. W., Omri, A., and Patel, G. B. (2001). *PCT International Publication No.* WO01/26683A2.
65. Lipford, G. B., Lehn, N., Bauer, S., Heeg, K., and Wagner, H. (1994). *Immunol Lett* **40,** 101-104.
66. Moore, M. W., Carbone, F. R., and Bevan, M. J. (1988). *Cell* **54,** 777-785.

Chapter 3

SOLID LIPID NANOPARTICLES

Anne Saupe and Thomas Rades
School of Pharmacy, University of Otago, P.O. Box 56, Dunedin, New Zealand

Abstract: In the last decade of the last century, solid lipid nanoparticles (SLN) have been introduced to the literature as a novel carrier system for cosmetic active ingredients and pharmaceutical drugs. SLN consist of biodegradable physiological lipids or lipidic substances and stabilisers which are generally recognised as safe (GRAS) or have a regulatory accepted status. Compared to other delivery systems such as liposomes, microemulsions and polymeric nanoparticles, SLN possess various advantages. Examples are production without organic solvents, long time physical stability and the possibility of protection of chemically labile actives inside the particles. This chapter describes the composition, production procedures, characterisation of quality and stability of SLN dispersions and describes possibilities for application of this colloidal carrier system.

Key words: lipid nanoparticles, production methods, physical stability, colloidal carriers, cosmetic application

1.   INTRODUCTION

In recent years, biocompatible lipid micro- and nanoparticles have been reported as potential drug carrier systems [1]. Colloidal carriers offer many advantages as delivery systems, such as increased bioavailability for poorly water soluble drugs [2]. Beside nanoemulsions, nanosuspensions, mixed micelles, liposomes and polymeric nanoparticles, melt-emulsified nanoparticles based on lipids solid at room temperature have been developed. Several problems of polymer based nanoparticles systems need to be overcome. These include residues from the organic solvents used in the production process [3], polymer toxicity and the scaling up of the production

M.R. Mozafari (ed.), Nanocarrier Technologies: Frontiers of Nanotherapy, 41–50.

process to industrial levels. The following positive features emphasise the potential use of solid lipid particles as drug carrier systems:

- A wide potential application spectrum (dermal, per os, intravenous)

- High pressure homogenisation as an established production method

- Avoidance of organic solvents

- Solid matrix composition of physiological and well-tolerated lipids

- Ability to produce and sterilise formulations on an industrial scale

2.   DEFINITION AND GENERAL INGREDIENTS

The first patents based on SLN formulations were issued in 1993 and 1996 [4]. Through the work of different research groups, the SLN delivery systems have been intensively investigated with respect to production, characterisation and application [5]. By definition, SLN-formulations are made from solid lipids (i.e. lipids solid at room temperature), surfactant(s) and water. The term lipid includes triglycerides, partial glycerides, fatty acids, steroids and waxes. A potential advantage of SLN is the use of physiological lipids which have low cytotoxicicity and therefore a decreased danger of acute and chronic toxicity [6, 7]. Examples of such lipids are triglycerides including tricaprin, trilaurin, tripalmitin, hard fat types including glycerol behenate and Glycerol palmitostearate, and waxes such as cetyl palmitate. Many emulsifiers, differing with respect to charge and molecular weight, have been used to stabilise the aqueous lipid dispersion. The choice of the surfactant and its concentration depends on the chosen lipid and on the administration route and is more limited for parenteral administration than for oral or dermal application. Examples of such surfactants include soybean lecithin, egg lecithin, Poloxamer 188 and sodium cholate.

3.   PRODUCTION PROCEDURES

Different methodologies exist for the production of finely dispersed lipid nanoparticle dispersions. The following section discusses the various methods currently available for the production of SLN, with focus on the ability to scale up the process to an industrial level.

3.1   High pressure homogenisation

High pressure homogenisation (HPH) has emerged as a reliable and powerful technique for the production of nanoemulsions for parenteral nutrition [5]. Scaling up is unproblematic in most cases. HHP can be performed at elevated temperatures (hot HPH) or at or below room temperature (cold HPH). High pressure homogenisers force a liquid at high pressure (100-2000 bar) through a narrow gap (few micrometers). The particle size is decreased by very high shear stress and cavitation forces. For HPH, the lipid and drug are melted (approximately 5-10°C above the melting point of the lipid). Typical lipid concentrations are in the range 5-20%. The aqueous phase at the same temperature as the lipid and containing a surfactant is then added to the lipid phase and a hot pre-emulsion is formed by high speed stirring (Fig. 3-1).

The hot pre-emulsion is then processed in a temperature controlled high pressure homogeniser, where generally 3 cycles at 500 bar are performed (APV LAB 40, Fig. 3-2, left). The obtained nanoemulsion solidifies by cooling down to room temperature. High pressure homogenisers are characterised by a homogenous power distribution due to the small size of the homogenising gap (25-30 μm) leading to particle sizes of approximately 200 nm and a narrow particle size distribution. The Gaulin 5.5 (Fig. 3-2, right) is also a piston gap homogeniser used for the large scale production of different formulations of up to 50 kg [8]. The LAB 40 (Fig. 3-2, left) differs from the Gaulin 5.5 in terms of thermoregulation, maximum production capacity, maximum pressure and the number of valves and pistons. Production using a Gaulin 5.5 is possible in a discontinuous and a circulating mode. For the discontinuous mode the formulation is collected in the product container, and after completing the first cycle the entire product is fed back to the feeding container for a repeat cycle. This is continued until the required number of cycles has been performed. The circulating mode is characterised by product feed back directly via a temperature controlled double walled tube to the feeding container. The variable production parameters include production time and homogenisation pressure.

Cold HPH is used for temperature labile- or hydrophilic drugs. First a suspension is obtained by melting lipid and drug together followed by rapid grinding under liquid nitrogen forming solid lipid microparticles. The homogenisation is carried out with the solid lipid including drug, i.e. a high pressure milling of a suspension. The homogenisation conditions are generally five cycles at 500 bar.

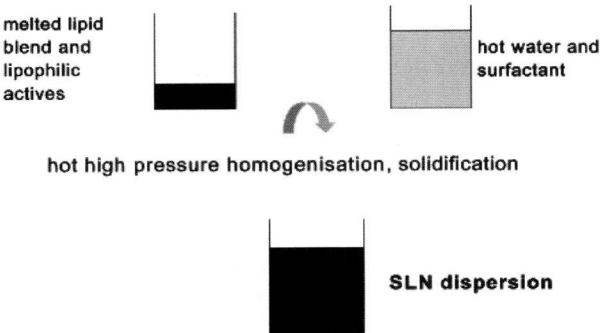

Figure 3-1. SLN production by hot high pressure homogenisation.

Figure 3-2. APV LAB 40 for batch sizes up to 40 mL (left) and Gaulin 5.5 in series with feeding container (150 kg / hour) (right).

3.2   Microemulsion based SLN preparations

This method is based on the dilution of microemulsions containing a molten lipid, surfactant, cosurfactant and water [9]. The hot microemulsion is then dispersed in cold water (2-3°C) under stirring and the excess water removed by ultra-filtration or lyophilisation in order to increase the particles concentration. Due to the dilution step, the achievable lipid contents are considerably lower when compared to HPH based formulations. Furthermore, high concentrations of surfactants and cosurfactants are required for the production process, which is undesirable with respect to regulatory requirements.

3.3   Preparation by solvent emulsification

This production method is based on precipitation of lipids in oil/water emulsions [10]. The lipid phase is dissolved in a water-immiscible organic solvent and is than emulsified in an aqueous phase before evaporation of the solvent under reduced pressure. The lipid precipitates forming SLN. An advantage is the avoidance of thermal stress, which makes this method suitable for thermolabile drugs. Disadvantages are solvent residues in the final dispersion and that these dispersions are generally quite dilute, due to the limited solubility of the lipid in the organic material.

4.   CHARACTERISATION OF QUALITY AND STABILITY OF SLN DISPERSIONS

An adequate characterisation of solid lipid nanodispersions is a necessity for the control of the quality of the product. However such characterisation is a serious challenge due to the colloidal size of the particles and the complexity of the system [11, 12]. The stability of SLN dispersions has been investigated intensively, e.g. determination of the particles size (photon correlation spectroscopy, PCS, laser diffraction, LD, electron microscopy), charge (Zeta potential, ZP) and thermal analysis (differential scanning calorimetry, DSC) [13, 14].

PCS is a good tool for characterisation of nanoparticles, but is not suitable for the detection of larger microparticles. The latter can be detected by LD measurements. Electron microscopy provides additional direct information on the particle shape [15]. The physical stability of optimised SLN dispersions is generally more than 12 months [13]. The average diameter of the main particle population remained stable between 200 and 230 nm for the investigated period of time. ZP measurements allow predictions about the storage stability of colloidal dispersions. Generally, dispersions with a ZP higher than ±30 mV are considered as physically stable due to electrostatic repulsion. Measurements of crystallinity and lipid modification are necessary because these parameters are strongly correlated with drug incorporation and release rates [16]. DSC is widely used to investigate the solid state of the lipid. The basis for DSC is the fact that different lipid modifications possess different melting points and melting enthalpies. SLN dispersions for parenteral application need to be sterilised. Various research groups have published particle size and ZP data of sterilised SLN dispersions revealing stability of the formulations. In one study SLN were sterilized by autoclaving. The formulations were stable during sterilization and the SLN maintained a spherical shape and narrow size distribution as confirmed by TEM analysis. SLN dispersions in water did not present particles larger than 1 μm after storage at 4°C for 1 year; they were freeze-dried after sterilization to obtain dry products [17].

5.   APPLICATION AND ADMINISTRATION ROUTES

5.1   Cosmetic application

An area of potentially lucrative application are topical cosmetic products based on SLN technology [18]. SLN are composed of well-tolerated excipients and due to their small particle size they possess adhesive properties leading to film formation on the skin. Further it has been shown that SLN have occlusive properties in vitro depending on their size, crystalline status and lipid concentration. This is due to the fact that SLN form an intact film upon drying and thus decrease water evaporation from the skin to the atmosphere [19]. SLN are physically stable in aqueous dispersions and also after incorporation into a dermal cream as shown by size measurements and DSC [20]. The occlusion promotes the penetration of vitamin E into the skin, as shown by a stripping test [19]. SLN loaded with vitamin A (retinol and retinyl palmitate) and incorporated in a hydrogel and o/w-cream were tested with respect to their influence on drug penetration into porcine skin [21]. Conventional formulations served as comparison. The best results were obtained with retinol SLN incorporated in the oil-in-water (o/w) cream retarding drug expulsion. The penetration of the occlusion sensitive drug retinyl palmitate was even more influenced by SLN incorporation [21]. SLN were also established as an UV protection system [20]. The particles themselves may act as UV blocker due to their particulate character. The incorporation of molecular sunscreens into the matrix of the particles leads to a synergistic effect of molecular sunscreen and the UV scattering. This means that the total content of molecular sunscreens in the formulation can be reduced while maintaining the protection level when compared to emulsions with molecular sunscreens [22]. In general the incorporation of sunscreens into lipid carriers improves the UV blocking effect whereas the increase at the different wavelengths depends also on the nature of the lipid matrix. High crystallinity of the SLN improves the UV blocking effect.

5.2 Peroral and parenteral administration

Formulations for oral administration forms of SLN may include aqueous dispersions or SLN loaded traditional dosage forms, e.g. tablets, pellets or capsules [23]. Studies were carried out for orally administered camptothecin (CA)-loaded nanoparticles. The encapsulation efficiency of camptothecin was 99.6%, and *in vitro* drug release was achieved for a period of seven days. In tested organs, the AUC (area under concentration/time curve) and mean residence time of CA-SLN increased significantly compared with a CA-control solution, and the increase in brain AUC was the highest among all tested organs [24]. Solid lipid nanoparticle delivery systems enhanced diterpenoid triepoxid (TP) absorption, reported to be effective in the treatment of patients with a variety of inflammatory and autoimmune diseases especially rheumatoid arthritis), increasing the bioavailability. TP-SLN increased the anti-inflammatory activity, protecting the cells from oxidative stress, lipid peroxidation processes and DNA damage [25] .

The parenteral application for SLN ranges from intraarticular to intravenous administration. The pharmacokinetics of doxorubicin incorporated as an ion-pair into SLN was compared with that of a commercial solution of the drug. The anthracycline concentration in the blood was markedly higher at each point time with the SLN formulation than with the commercial solution. The drug concentration was also higher in the lung, spleen and brain. SLN-treated rats demonstrated a lower doxorubicin concentration in liver, heart and kidney. The results indicate that SLN increased the area under the curve (0–180 min) of doxorubicin when compared to conventional doxorubicin solution and led to a different body distribution profile for the drug [26].

5.3 The application of SLN as a vaccine adjuvant

Adjuvants are used in vaccination to enhance the immune response. Subunit vaccines have improved safety over traditional vaccines. They are less effective in inducing protective immune responses and therefore effective adjuvants are required. Due to the lipidic nature of SLN vaccine formulations containing such nanoparticles will be degraded more slowly (due to the solid state of the carrier) providing a longer lasting exposure to the immune system with persistence of the vaccine antigen. Immunisation studies in chicken were performed by using a mycoplasma bovis antigen and mouse immunoglobulin G. The resulting antibodies were extracted from egg yolk. The adjuvant effect was compared to freund's complete/incomplete adjuvant (FCA/FIA) and to the vaccine without any adjuvant. FCA revealed the highest antibody titer against mouse IgG. SLN with a particle size of more than 100 nm exhibited a clear adjuvant activity, whereas SLN smaller than 100 nm, revealed a lower adjuvant acitivity. The SLN induced characteristic changes in the chronological titer development, suggesting some adjuvant effect. SLN formulations were well tolerated with no significant tissue toxicity observed [27].

6.   SUMMARY AND OUTLOOK

SLN are a novel and innovative therapeutic delivery system. Clear advantages include the composition (physiological compounds), the effective production process (especially the possibility of large scale production), avoidance of organic solvents during the production process and the possibility of producing highly concentrated lipid dispersions. SLN are a complex system due to the physical state of the lipid. Appropriate characterisation of the formulation requires several analytical methods and will decide over the SLN administration and the most suitable application route. Results obtained with dermal formulations are currently the most promising and thus will probably be the main application for SLN in the future.

7.   REFERENCES

1. Barratt, G.M. *Therapeutic applications of colloidal drug carriers.* Pharmaceutical Science & Technology Today, 2000. **3**(5): 163-171.
2. Barratt, G. *Colloidal drug carriers: achievements and perspectives.* Cellular and Molecular Life Sciences (CMLS), 2003. **60**(1): 21-37.
3. Zweers, M.L., Engbers, G.H., Grijpma, D.W. and Feijen, J. *In vitro degradation of nanoparticles prepared from polymers based on DL-lactide, glycolide and poly(ethylene oxide).* Journal of Controlled Release, 2004. **100**(3): 347-356.

4.   Muller, R.H., Maassen, S., Weyhers, H., Specht, F. and Lucks, J.S. *Cytotoxicity of magnetite-loaded polylactide, polylactide/glycolide particles and solid lipid nanoparticles.* International Journal of Pharmaceutics, 1996. **138**(1): 85-94.

5.   Schwarz, C., Mehnert, W., Lucks, J.S. and Muller, R.H. *Solid lipid nanoparticles (SLN) for controlled drug delivery. I. Production, characterization and sterilization.* Journal of Controlled Release, 1994. **30**(1): 83-96.

6.   Muller, R.H., Ruhl, D., Runge, S., SchulzeForster, K. and Mehnert, W. *Cytotoxicity of Solid Lipid Nanoparticles as a Function of the Lipid Matrix and the Surfactant.* Pharmaceutical Research, 1997. **14**(4): 458-462.

7.   Heydenreich, A.V., Westmeier, R., Pedersen, N., Poulsen, H.S. and Kristensen, H.G. *Preparation and purification of cationic solid lipid nanospheres–effects on particle size, physical stability and cell toxicity.* International Journal of Pharmaceutics, 2003. **254**(1): 83-87.

8.   Gohla, S.H. and Dingler, A. *Scaling up feasibility of the production of solid lipid nanoparticles (SLN).* Die Pharmazie, 2001. **56**(1): 61-63.

9.   Ugazio, E., Cavalli, R. and Gasco, M.R. *Incorporation of cyclosporin A in solid lipid nanoparticles (SLN).* International Journal of Pharmaceutics, 2002. **241**(2): 341-344.

10. Trotta, M., Debernardi, F. and Caputo, O. *Preparation of solid lipid nanoparticles by a solvent emulsification-diffusion technique.* International Journal of Pharmaceutics, 2003. **257**(1-2): 153-160.

11. Venkateswarlu, V. and Manjunath, K. *Preparation, characterization and in vitro release kinetics of clozapine solid lipid nanoparticles.* Journal of Controlled Release, 2004. **95**(3): 627-638.

12. Lim, S.J. and Kim, C.K. *Formulation parameters determining the physicochemical characteristics of solid lipid nanoparticles loaded with all-trans retinoic acid.* International Journal of Pharmaceutics, 2002. **243**(1-2): 135-146.

13. Mehnert, W. and Mader, K. *Solid lipid nanoparticles: Production, characterization and applications.* Advanced Drug Delivery Reviews, 2001. **47**(2-3): 165-196.

14. Bunjes, H., Westesen, K. and Koch, M.H.J. *Crystallization tendency and polymorphic transitions in triglyceride nanoparticles.* International Journal of Pharmaceutics, 1996. **129**(1-2): 159-173.

15. Muehlen, A.Z., Muehlen, E.Z., Niehus, H. and Mehnert, W. *Atomic Force Microscopy Studies of Solid Lipid Nanoparticles.* Pharmaceutical Research, 1996. **13**(9): 1411-1416.

16. Westesen, K., Bunjes, H. and Koch, M.H.J. *Physicochemical characterization of lipid nanoparticles and evaluation of their drug loading capacity and sustained release potential.* Journal of Controlled Release, 1997. **48**(2-3): 223-236.

17. Cavalli, R., Caputo, O., Carlotti, M.E., Trotta, M., Scarnecchia, C. and Gasco, M.R. *Sterilization and freeze-drying of drug-free and drug-loaded solid lipid nanoparticles.* International Journal of Pharmaceutics, 1997. **148**(1): 47-54.

18. Lippacher, A., Muller, R.H. and Mader, K. *Semisolid SLN(TM) dispersions for topical application: influence of formulation and production parameters on viscoelastic*

properties. European Journal of Pharmaceutics and Biopharmaceutics, 2002. **53**(2): 155-160.

19. Wissing, S.A. and Muller, R.H. *The influence of solid lipid nanoparticles on skin hydration and viscoelasticity - in vivo study.* European Journal of Pharmaceutics and Biopharmaceutics, 2003. **56**(1): 67-72.

20. Wissing, S.A. and Muller, R.H. *Cosmetic applications for solid lipid nanoparticles (SLN).* International Journal of Pharmaceutics, 2003. **254**(1): 65-68.

21. Jenning, V., Gysler, A., Schafer-Korting, M. and Gohla, S.H. *Vitamin A loaded solid lipid nanoparticles for topical use: occlusive properties and drug targeting to the upper skin.* European Journal of Pharmaceutics and Biopharmaceutics, 2000. **49**(3): 211-218.

22. Saupe, A., *Pharmaceutical and cosmetic applications of nanostructured lipid carriers (NLC): sun protection and care.* 2004, Free University of Berlin.

23. Pouton, C.W., *Lipid formulations for oral administration of drugs: non-emulsifying, self-emulsifying and 'self-microemulsifying' drug delivery systems.* European Journal of Pharmaceutical Sciences, 2000. **11**(Supplement 2): S93-S98.

24. Yang, S., Zhu, J., Lu, Y., Liang, B. and Yang, C. *Body Distribution of Camptothecin Solid Lipid Nanoparticles After Oral Administration.* Pharmaceutical Research, 1999. **16**(5): 751-757.

25. Mei, Z., Li, X., Wu, Q., Hu, S. and Yang, X. *The research on the anti-inflammatory activity and hepatotoxicity of triptolide-loaded solid lipid nanoparticle.* Pharmacological Research, 2005. **51**(4): 345-351.

26. Zara, G.P., Cavalli, R., Fundaro, A., Bargoni, A., Caputo, O. and Gasco, M.R. *Pharmacokinetics of Doxorubicin incorporated in solid lipid nanosheres (SLN).* Pharmacological Research, 1999. **40**(3): 281-286.

27. Olbrich, C., Mueller, R.H., Tabatt, K., Kayser, O., Schulze, C. and Schade, R. *Stable biocompatible adjuvants--a new type of adjuvant based on solid lipid nanoparticles: a study on cytotoxicity, compatibility and efficacy in chicken.* Altern. Lab. Anim., 2002. **30**(4): 443-58.

Chapter 4

HYDROTROPIC NANOCARRIERS FOR POORLY SOLUBLE DRUGS

Tooru Ooya[1], Sang Cheon Lee[2], Kang Moo Huh[3] and Kinam Park[4]

[1]School of Materials Science, Japan Advanced Institute of Science and Technology, 1-1 Asahidai, Tatsunokuchi, Ishikawa, 923-1292, Japan; [2]Korea Institute of Ceramic Engineering and Technology, 233-5 Gasan-dong, Guemcheon-gu, Seoul 153-801, South Korea; [3]Chungnam National University, Dept. of Polymer Science and Engineering, Yuseong-gu Gung-dong 220, Daejeon, 305-764, South Korea; [4]Departments of Pharmaceutics and Biomedical Engineering, Purdue University, West Lafayette, IN 47907-1336, U.S.A.

Abstract: Delivery of poorly water-soluble drugs remains as one of the most difficult challenges in the pharmaceutics and drug delivery areas. One of the recent approaches of increasing the water-solubility of poorly soluble drugs has been to utilize polymeric hydrotropic agents. Hydrotropic agents in nanocarrier forms, such as dendrimers and polymer micelles, increase the water solubility by orders of magnitude. The hydrotropic nanocarriers have advantages over other carriers for their high drug loading efficiency and long-term stability.

Key words: hydrotropy, dendrimer, polymer micelle, water-solubility, paclitaxel, nanocarrier

1. INTRODUCTION

Recent advances in nanoscience and nanotechnology provide us with a great opportunity to understand biological systems in nano-scale, and such understanding has created new research fields such as "bio-nanotechnology" [1]. New designs of nanocarriers based on structural characteristics in nano-scale are expected, because the interdisciplinary fields of bio-nanotechnology undoubtedly have a great potential of new bio-functions toward regulation of biological systems. It is known that biological processes including large macromolecular self-assemblies are based on dynamic consequences of cooperativity and metastability. The assembly of large supramolecular

51

M.R. Mozafari (ed.), Nanocarrier Technologies: Frontiers of Nanotherapy, 51–73.
© 2006 Springer. Printed in the Netherlands.

complexes produces dynamic features in a construct. Supramolecular architectures formed via non-covalent interactions are expected to play an important role in terms of dynamic functions under physiological conditions. Generally, self-assembly, such as protein folding, involves hydrophobic interaction [2], existing in water at hydrophobic interface of the systems [3]. Such self-assembly via hydrophobic interaction plays an important role for creating nanocarriers that deliver poorly water-soluble drugs [4-6].

Nanocarriers are highly useful for delivering poorly soluble drugs without drug solubilizers that may be toxic and for targeting to suitable tissues and cells. There have been various approaches for increasing the aqueous solubility of poorly soluble drugs via self-assembly. Polymeric micelles have been extensively studied as a promising drug formulation that can effectively dissolve various types of hydrophobic drugs with high drug loading capacity for increased bioavailability. Hydrophobic drugs can be dissolved or physically entrapped in the core of polymeric micelles at concentrations that can exceed their intrinsic water-solubility by orders of magnitude. Recently, our research groups have introduced new polymeric systems known as the hydrotropic polymer, hydrotropic dendrimer, and hydrotropic polymer micelle. They can increase the solubility of poorly soluble drugs by several orders of magnitude. The monomers of those hydrotropic polymers were designed based on the molecular structures of low molecular weight hydrotropic agents (or hydrotropes) which are highly effective in solubilizing poorly soluble drugs. In the case of hydrotropic dendrimers, increased density of drug-solubilizing molecules in a dendrimer contributes to the effective solubilization. The hydrotropic polymers and dendrimers have been developed into hydrotropic polymer micelles that can act as nanocarries for poorly water soluble drugs. This chapter introduces hydrotropic nanocarries based on hydrotropic polymers and dendrimers for delivery of poorly soluble drugs.

2.   WHAT IS HYDROTROPIC SOLUBILIZATION?

Hydrotropy refers to an increase in water solubility caused by the addition of large amount of a second solute [7]. However, the exact meaning of hydrotropy is still unclear. There have been various theoretical and experimental studies aiming at explanation of hydrotropic solubilization [8, 9]. Most of hydrotropic molecules are composed of an aromatic ring substituted by anionic or cationic moieties. The hydrotropic molecules are usually too small to induce micelle formation. The hydrotropic molecules aggregate at a certain concentration, known as the minimum hydrotropic concentration (MHC), to increase the solubility of poorly soluble drugs significantly [10]. For example, nicotinamide has been shown to enhance the solubilities of a wide variety of hydrophobic drugs through complexation [11-13]. Here, the aromaticity of the pyridine ring of nicotinamide and its derivatives, such as *N*-methylnicotinamide and *N,N*-diethylnicotinamide, promotes stacking of molecules through its planarity for the aggregation [12]. Another interesting phenomenon of hydrotropic solubilization is self-association of the hydrotrope in an aqueous phase. Some experimental data indicate that some hydrotropes, including nicotinamide and aromatic sulfonates, associate in aqueous solutions [7]. Studies on the nicotinamide–riboflavin system showed that the self-association of nicotinamide contributed to the solubility increase of riboflavin rather than complexation between two species [7]. According to the report of Silva et al., the micellization process does not occur in hydrotrope-water mixtures, and the the MHC of usual hydrotropes is about 1 M, which is higher than typical critical micelle concentrations (CMC) of 10^{-2}–10^{-3} M [9]. Thus, both high concentration and association are needed to induce hydrotropic solubilization.

Hydrotropes solubilize hydrophobic drugs, but it is difficult to find out the structure-activity relationship for the hydrotropic solubilization. Hydrotropic solubilization of paclitaxel has been studied using more than 60 candidate hydrotropic agents and their analogues [14]. Several effective hydrotropic structures were identified for their ability to solubilize paclitaxel, and some of them are shown in Fig. 4-1. Among them, *N,N*-diethylnicotinamide (DENA) was found to be the most effective hydrotropic agent for paclitaxel. The solubility of paclitaxel was 5~6 orders of magnitude greater than the intrinsic solubility of paclitaxel. *N*-Picolylnicotinamide (PNA), *N*-allylnicotinamide, and sodium salicylate were also found to have high solubilizing capacity for paclitaxel. This information can be used to find other hydrotropic compounds and to design new hydrotropic analogues that are effective for paclitaxel and other poorly soluble drugs.

3. DESIGN OF HYDROTROPIC POLYMERS

3.1 Hydrotropic polymers

One barrier for successful oral drug delivery using hydrotopes is the co-absorption of a significant amount of low molecular weight hydrotropes along with the drug to be delivered. Thus, there was a need for developing polymeric form of hydrotropes effective in increasing the drug solubility. The polymeric forms of the hydrotropic agents may provide various advantages, including prevention of co-absorption from the GI tract after oral administration while maintaining the beneficial hydrotropic properties.

The polymers used to increase the water solubility of hydrophobic compounds were found in the patent literature. The structures of those polymers were based on polyvinylpyrrolidone (PVP). The investigators synthesized new pyrrolidone-containing polymers of which side groups had considerably reduced steric hindrance when complexing with water insoluble organic compounds [15, 16]. These polymers increased the water solubility of poorly soluble drugs, such as furosemide, indomethacin, and triamterene. The chemical structures of these polymers were based only on PVP, and thus the structural variation of the polymer has been limited. Therefore, it is important to systematically diversify the structure of the hydrotropic polymers so that their property can be tailored for specific drug solubilization.

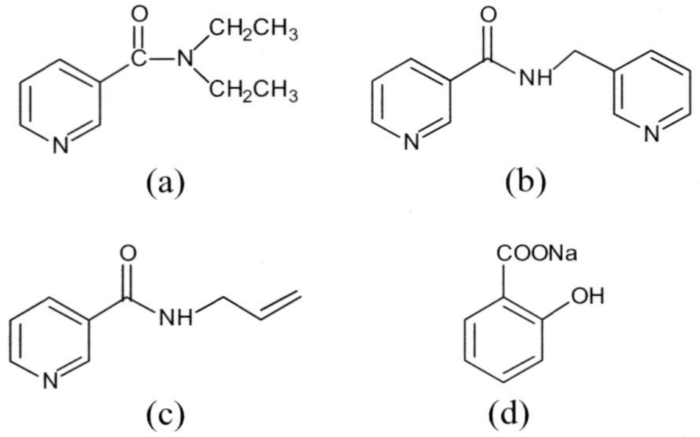

Figure 4-1. Several hydrotropic agents identified for paclitaxel solubilization: N,N-diethylnicotinamide (a), N-picolylnicotinamide (b), N-allylnicotinamide (c), and sodium salicylate (d).

3.2   Design parameters of hydrotropic polymers and structure-property relationship

Since the exact mechanisms of solubility increase of poorly soluble drugs by hydrotropes are not known, it is difficult to predict the structural requirements of hydrotropes for effective solubilization of various drugs. Furthermore, when the concept of hydrotropic solubilization is extended from simple low molecular weight species to the polymeric forms, the structural needs for the hydrotropy are still unpredictable and become more complicated. Thus, the most rational approach for the synthesis of hydrotropic polymers is utilizing the chemical structures of the most promising low molecular weight hydrotropic agents. Hydrotropic properties of various low molecular weight molecules were examined for their ability to increase water-solubility of paclitaxel [14]. From the study, the information on structure-activity relationship and structural requirements of the hydrotropes for the paclitaxel solubilization were obtained. Hydrotropic polymers were designed based on the excellent hydrotropes, such as *N*-picolylnicotinamide (PNA) and *N,N*-diethylnicotinamide (DENA). PNA and DENA at the 3.5 M concentration increased the water solubility of paclitaxel up to 30 and 40 mg/mL, respectively.

To synthesize hydrotropic polymers, several design parameters should be considered. They are the polymer backbone, the type of hydrotropes, orientation of hydrotropic moieties to the polymer backbone, and spacer groups bridging the polymer backbone and the hydrotropic moieties. The first step for hydrotropic polymers is the modification of the hydrotropes with polymerizable groups. This step has a significant meaning since it dictates all subsequent design parameters. Estimation of hydrotropic property of modified hydrotropic agents is important since the type of the spacer group between the polymer backbone and the hydrotropic agent may significantly affect their hydrotropic property. Enhancement of aqueous solubility of paclitaxel was observed with increasing the concentration of 2-(4-vinylbenzyloxy)-*N*-picolylnicotinamide (2-VBOPNA), 6-(4-vinylbenzyloxy) -*N*-picolylnicotinamide (6-VBOPNA), and PNA [17]. It is noteworthy that 2-VBOPNA and 6-VBOPNA, with a vinylbenzyloxy group linked to 2- and 6-position of pyridine ring, showed rather different pattern of solubilization but retained comparable hydrotropic property for the paclitaxel to that of PNA. 2-VBOPNA enhanced the aqueous solubility of paclitaxel to a larger extent than 6-VBOPNA did at a wide concentration range. This finding suggests that the linked orientation of hydrotropic moiety (e.g. PNA) to the polymerizable group is the one of the key factors deciding the hydrotropic property of the modified hydrotropes. Even though the hydrotropic moieties can be linked into the commercially available polymer backbone, the design of the polymerizable hydrotropic agent is more promising in generation of excellent hydrotropic polymers. For hydrotropic polymers to efficiently increase water-solubility of drugs, the high content of hydrotropic moieties in the polymer is essential since the hydrotropic action is presented only in the high local concentration of hydrotropic groups. The simple conjugation of the hydrotropic moieties into the polymer backbone does not guarantee the high degree of hydrotropes in the polymer structure. On the other hand, the polymerization of the hydrotropes modified with polymerizable groups can produce the polymers with hydrotropic moieties in every repeating unit, thereby maximizing the content of the hydrotropic structure in the polymer chains. Various polymer backbones can be obtained by introducing allyloxy, acryloyl, styryl, and oligoethyleneoxyacryloyl groups to the hydrotropic moieties. PNA was used as a hydrotropic moiety for the synthesis of hydrotropic polymers with the styryl backbone [18]. High molecular weight polymers could be synthesized by radical polymerization, and the polymers had the high content of PNA groups. Fig. 4-2 shows the synthetic app-roach for poly (2-(4-vinylbenzyloxy)-*N*-picolylnicotinamide) P(2-VBOPNA) and Poly (2-(2-(acryloyloxy) ethoxyethoxyethoxy)-*N*-picolylnicotinamide) 'P(ACEEEPNA)'.

Effect of the design parameters on the hydrotropic properties of the polymers was examined [18]. The spacer type and the binding orientation of the hydrotropic moieties to the polymer backbone significantly affected the solubility-enhancing properties for paclitaxel. Fig. 4-3 shows the solubility increase of paclitaxel as a function of PNA and PNA-containing hydrotropic polymers. P(ACEEEPNA) retained the hydrotropic property of PNA to increase the water-solubility of paclitaxel to 0.32 mg/mL at the maximum polymer concentration of 290 mg/mL. As compared with the intrinsic water-solubility of paclitaxel (0.3 µg/ml), this is 1000-fold increase in solubility. In contrast to PNA, P(ACEEEPNA) solubilized paclitaxel to a significant extent even at the low concentration range, e.g. less than 50 mg/mL. The polymer solubilized paclitaxel even at the very low concentration range, where solubilization of paclitaxel by PNA was not observed. P(2-VBOPNA) and P(6-VBOPNA) also retained the hydrotropic property of PNA and increased the water-solubility of paclitaxel to 0.56 mg/mL and 0.13 mg/mL respectively. There was a big difference in the hydrotropic property between the polymer having an oligo(ethylene glycol) spacer and the polymers with a phenyl spacer. The hydrotropic property of P(2-VBOPNA) and P(6-VBOPNA) was much more pronounced than that of P(ACEEEPNA). P(ACEEEPNA) showed a positive curvature as its concentration increased, while P(2-VBOPNA) and P(6-VBOPNA) showed negative curvatures in the aqueous paclitaxel solubility as a function of the polymer concentration. It is interesting to notice that P(2-VBOPNA) is a much better hydrotropic polymer than P(6-VBOPNA). Apparently, the orientation of the PNA moiety in P(2-VBOPNA) has better stacking ability to solubilize paclitaxel. It appears that the PNA moieties bound to the flexible oligo(ethylene glycol) spacer are favored to form high orders of associated structures with increasing concentration. Thus, the structure and property of a spacer group between the polymer backbone and the hydrotropic moiety play a key role in modulating the solubilization profile by polymeric hydrotropes. Hydrotropic polymers can be made using the same hydrotropic moiety but with different orientations by copolymerization of monomers obtained from the same hydrotrope. This approach can provide an opportunity of the facile interaction of hydrotropic units with paclitaxel by compensating the motional limitation of each polymer-bound hydrotropic moiety.

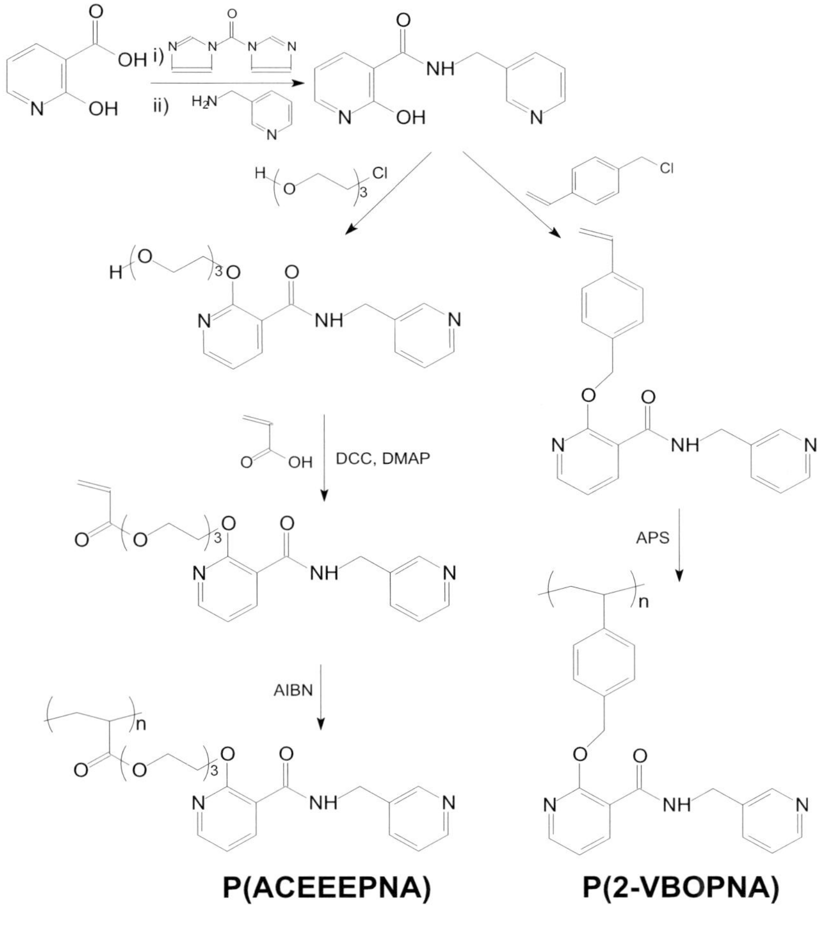

Figure 4-2. Synthetic route for P(ACEEEPNA) and P(2-VBOPNA).

Hydrotropic copolymers can also be made using two different hydrotropes as shown in Fig. 4-4. The concept of using two different hydrotropes on the same polymer backbone is based on the concept of the facilitated hydrotropy. The facilitated hydrotropy is the use of combination of different hydrotropic agents to yield higher hydrotropic property, compared to individual hydrotropes [10]. The maximum synergistic hydrotropic effect can be obtained by optimizing the factors, such as the type and length of spacers, orientations of the hydrotrope, and the use of different hydrotropes.

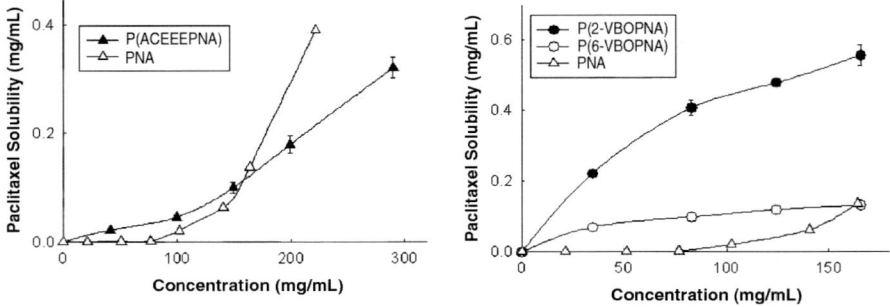

Figure 4-3. Enhancement of paclitaxel solubility as a function of PNA and PNA-containing hydrotropic polymers. (Modified from reference 18).

In general, the low molecular weight hydrotropes display efficient solubilizing ability beyond relatively high MHC values that range from 20 to 200 mg/mL or higher, depending on their classes [20]. Thus, it appears that the highly localized concentration of PNA moieties bound to the hydrotropic polymers facilitate self-association at a much lower concentration range. The lower MHC values, as compared with low molecular weight hydrotropic agents, might be one of the important characteristics of the hydrotropic polymers in that a small amount of the polymeric hydrotropes can self-associate and display the solubilizing ability for poorly water-soluble compounds. Based on the properties distinct to low molecular weight hydrotropes, one could define a hydrotropic polymer as a freely water-soluble polymer possessing the ability to increase aqueous solubility of poorly soluble compounds through association of its hydrotropic moieties. The appropriate selection and combination of the polymer backbone, the type of hydrotropes, binding orientation of hydrotropes, the spacer groups, and facilitated hydrotropic concept may result in the solubilizing systems of high potential for successful oral formulations of poorly water-soluble drugs.

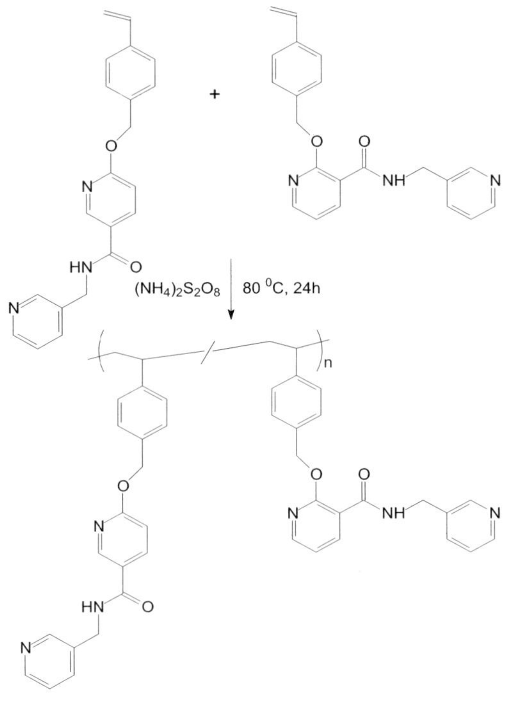

Figure 4-4. Synthetic scheme of P(2-VBOPNA-co-6-VBOPNA).

4. DESIGN OF HYDROTROPIC DENDRIMERS

4.1 Polyglycerol dendrimers as a hydrotrope

Dendrimers are nano-sized, highly branched macromolecules with monodispersed characters. Due to the nano-sized spherical shape and surface functionalities, chemical reactivity and physical properties, e.g. viscosity, are quite different from those of linear polymers [19]. Encapsulation of hydrophobic compounds in dendrimers has been extensively studied for drug delivery [20, 21]. In particular, biocompatible and/or biodegradable dendrimers have been used as drug delivery systems and tissue scaffolds [22,

23]. Polyglycerol dendrimers (PGDs) [24, 25] have a good potential as biomaterials because of high water solubility, chemical reactivity and structural similarity to poly(ethylene glycol) (PEG). So far, PEG has been used to modulate water solubility of poorly soluble drugs [26, 27]. PEG with molecular weight of 400 (PEG400) has been frequently used as a co-solvent to dissolve poorly water-soluble drugs [28, 29]. For example, PEG400 increased the aqueous solubility of ß-oestradiol by 4-5 orders of magnitude at its concentration of 80 wt% and higher [29]. It is suggested that the majority of PEG400 is believed to self-associate through hydrogen bonding mediated by water molecules at concentrations greater than 80 wt% [30]. From these observations, we hypothesized that PGD could act as a hydrotrope because the high density of ethylene glycol-like units in PGD can provide high local concentrations of PEG400.

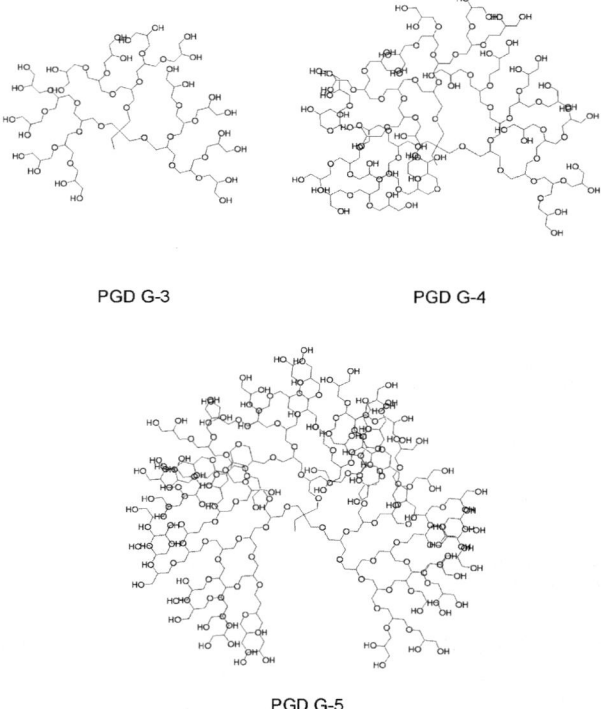

PGD G-3   PGD G-4

PGD G-5

Figure 4-5. Schematic structures of G-3, G-4 and G-5 polyglycerol dendrimers. (Modified from reference 32).

Polyglycerol dendrimers (PGDs) with 4-5 generations (Fig. 4-5) were synthesized and used to investigate the effect of dendritic architecture and its generation on aqueous solubilization of paclitaxel [31]. Chemical and physical properties of the PGDs were characterized by NMR, MALDI-TOF mass, GPC, viscosity and dynamic light scattering measurements (Table 4-1). The paclitaxel solubility in all solutions of PGDs, even below 10 wt%, was much higher than that in PEG400 (Fig. 4-6). Increase in the paclitaxel solubility by PGDs was dependent on the dendrimer generation. The dendritic structure was the reason for the enhanced solubility of paclitaxel even at low concentrations. ^{1}H NMR spectra of paclitaxel before and after mixing with PGDs in D_2O suggested that the aromatic rings and some methyne groups of paclitaxel were surrounded by PGDs (Fig. 4-7). PGDs, which do not require hydrophobic segment as in polymeric micelles, provide an alternative method of hydrotropic solubilization of poorly soluble drugs.

Table 4-1. Results of MALDI-MS, GPC, and Viscosity of PGDs.

Sample Code	M_{theo} [a]	MALDI	DLS	[η]
		M_w [b]	R_h (nm)	(mL·g^{-1})
PGD G-3	1,689	1,690	1.1	1.89
PGD G-4	3,508	3,507	1.9	3.90
PGD G-5	7,104	6,959	2.4	3.06

[a] Theoretical molecular weight, [b] Determined by MALDI-TOFF mass measurements

4.2   PEGylated hydrotropic dendrimers

In addition to the hydrotropic property of PGDs, surface modification of PGDs with PEG would provide even higher local concentrations of PEG. To evaluate the effect of PEG modification on hydrotropic solubilization, PGD was PEGylated by using methoxy-PEG with Mn of 550 and 2,000 [32]. PEGylated PGD with generations 4 (PEG-G4) was synthesized by conjugating *N,N'*-carbonyldiimidazole (CDI)-activated methoxy-PEG (Mn = 550, and 2,000) to PGD-G4. The CDI-activated methoxy-PEG was allowed to react with PGD-G4 in DMSO in the presence of dimethylaminopyridine (DMAP) and the solution was stirred at 70°C under N_2 for 24h to obtain PEGylated PGD-G4s, PEG500-PGD-G4 and PEG2000-PGD-G4 (Fig. 4-8).

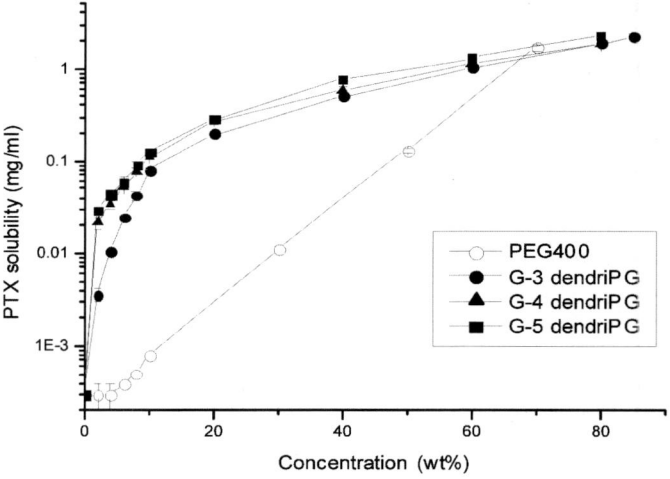

Figure 4-6. Aqueous paclitaxel (PTX) solubility as a function of the PGD concentration. (Modified from reference 31).

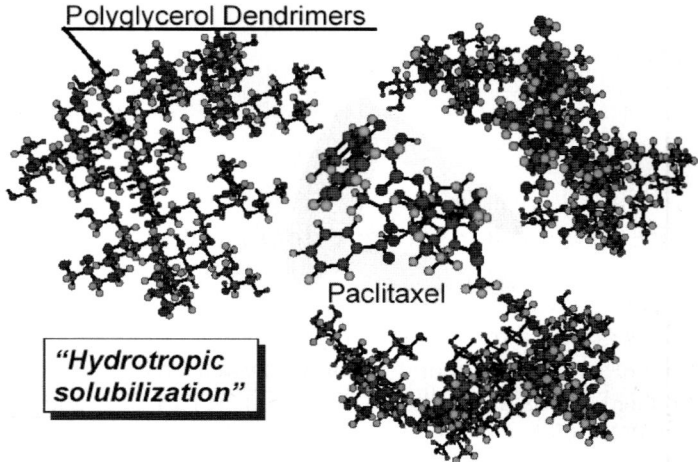

Figure 4-7. Image of hydrotropic solubilization by PGDs.

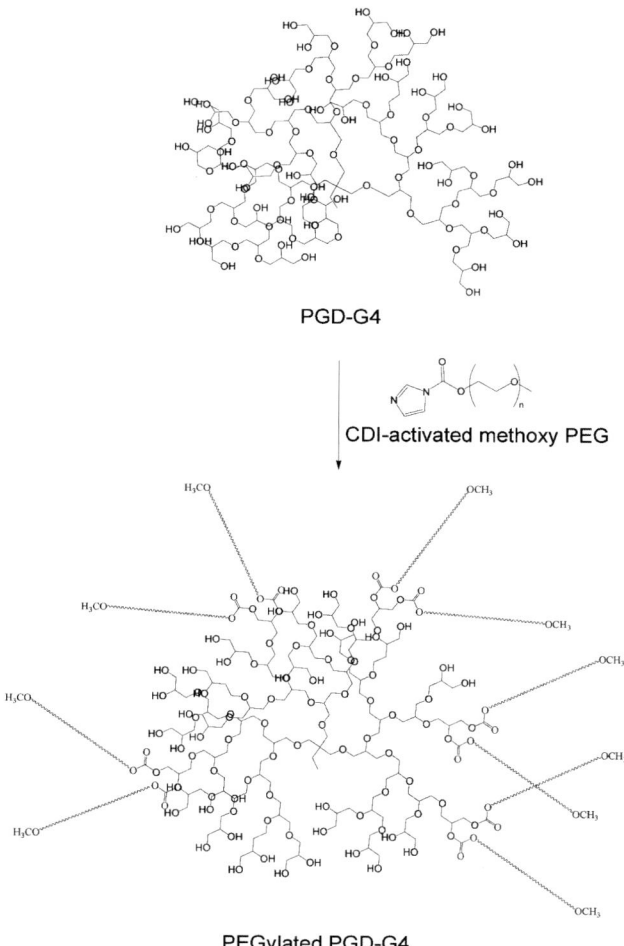

PGD-G4

CDI-activated methoxy PEG

PEGylated PGD-G4

Figure 4-8. Synthesis of PEGylated polyglycerol dendrimers (Modified from reference 42).

The ability of PGD G-4 to enhance the paclitaxel solubility was grater than that of PEG 400. PEG2000-PGD-G4 solubilizes paclitaxel at the similar level. On the other hand, PEG500-PGD-G4 showed the greatest enhancement: paclitaxel solubility was considerably increased over 2 wt%. These results suggest that the methoxy-PEG550 chains grafted on the surface of PGD-G4 contribute to interact with paclitaxel molecule. Methoxy-PEG-2000 was not effective due to the low solubility effect of PEG2000 [33]. Paclitaxel solubilities by PEGy500-PGD-G4 and PEGy2000-PGD-G4 at 10 wt% concentration each were 1,520-fold, and 408-fold higher, respectively, than the paclitaxel solubility in water (0.3 μg/ml) [14] (Fig. 4-9). This result suggests that increased local density of PEG550 due to conjugation to PGD surface provides a nano-scale space with a high concentration of PEG550.

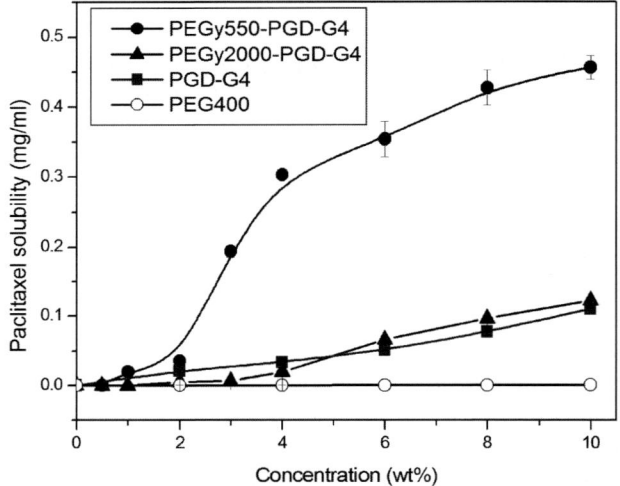

Figure 4-9. Palitaxel solubility as a function of concentration of PEG derivatives. (Modified from reference 42).

5.   SELF-ASSEMBLING SYSTEMS FOR DRUG SOLUBILIZATION

Surfactants are amphiphilic molecules that are characterized by distinct polar and nonpolar regions. In aqueous media, as the concentration of surfactant increases they self-assemble to form unique soluble structures called micelles. The concentration at which micelles begin to form is called the critical micelle concentration (CMC). These micelles are normally spherical and composed of a hydrophobic core and a hydrophilic shell. Nonpolar regions of the surfactant molecules are segregated from the aqueous exterior to form an inner core that is surrounded by a palisade of the polar regions which are in direct contact with the water.

There has been considerable interest in the interaction between hydrophobic solutes and micellar structures as a solubilization process by which solubility of the solutes can be enhanced [34]. A hydrophobic drug can be located within the micelle core during self-assembling process and becomes water-soluble as long as the micellar structures remain intact in aqueous media. There are a number of surfactants available for use as solubilizing agents. If the surfactant concentration falls below the CMC by dilution, the micellar structures can not be kept any more and a drug that was solubilized by the micelles precipitates from the aqueous medium [35]. In addition, several problems such as poor incorporation efficiency for many drugs, and thermodynamic instability, and toxicity have significantly limited their clinical applications. For these reasons, recently much interest has been placed on polymer micelles.

Amphiphilic block copolymers with diverse block structures may self-assemble in aqueous media to form micellar structures as low molecular weight surfactants do. It is generally known that polymeric amphiphiles have a much lower CMC than low molecular weight surfactants because they have more interaction sites. For this reason, micelles from amphiphilic block copolymers are more stable and so they have attracted a lot of attention as vehicles for solubilization and delivery of poorly water-soluble drugs [36-39]. They can act as water-soluble, biocompatible nano-carriers with a size of 10~100 nm and a proven efficacy of delivering poorly soluble drugs. Many advantages of using polymer micelles have been demonstrated and reviewed in the literature [37, 40, 41]. Polymer micelles with characteristic core-shell structures have successfully been used to solubilize poorly soluble drugs, such as paclitaxel [42], doxorubicin [43], cisplatin [44], amphotericin B [45], risperidone [46], and cyclosporine A [47].

5.1   Hydrotropic polymer micelles

In polymer micelle systems, the hydrophilic shell allows steric stability and thereby a long circulation of the drug, whereas the hydrophobic core are responsible for solubilizing and delivering poorly soluble drugs. Several major factors which influence the solubilizing property of polymer micelles have been studied [37]. They include the nature of the drug, nature of core-forming polymer block, block structure, and molecular weight. It has been demonstrated in many studies that the compatibility between the solubilizate (drug) and the core-forming polymer block is a critical factor to increase the loading capacity and loading efficiency. As a result, a polymer micelle system that can solubilize a given drug most effectively may be achieved by designing the hydrophobic polymer block that form micellar core with direct interaction with drug molecules.

Hydrotropic polymers that are synthesized by polymerization of monomers containing hydrotropic groups can be used as a hydrophobic building block for constructing micellar structures. There are a number of identified hydrotropic structures that are effective to enhance the water-solubility of various poorly soluble drugs. For example, nicotinamide derivatives were found to increase the water-solubility of paclitaxel by several orders of magnitude [14]. In addition, hydrotropic property was maintained in their polymeric forms (hydrotropic polymers) [18]. Hydrotropic polymers containing *N,N*-diethylnicotinamide (DENA) group in their repeating units was used as a core-forming polymer block [48]. Amphiphilic block copolymers could be synthesized by atom transfer radical polymerization of vinyl monomers containing DENA group using poly(ethylene glycol) macroinitiator as illustrated in Fig. 4-10. Hydrotropic polymer micelles, consisting of a hydrophilic poly(ethylene glycol) shell and a hydrophobic core that contains a significant amount of hydrotropic moieties, could be obtained via self-assembling property in aqueous media.

It is commonly observed that polymer micelles show low stability in drug-loaded state and the stability becomes even lower as the drug loading content increases. In many cases, the process for incorporation of drugs into micelles requires the use of organic solvents for dissolving either or both of the drug and the polymer because the micelle-forming polymer is not easily dissolved in water. On the other hand, based on synergistic effect of the unique micellar characteristics and hydrotropic activity, the hydrotropic polymer micelles exhibited a high drug loading capacity (up to 37.4 wt% for paclitaxel for example) with enhanced long-term stability. The poor physical stability of paclitaxel-loaded micelles was overcome by introducing hydrotropic moieties into the core structure of polymer micelle. Hydrotropic micelles could maintain their colloidal stability for more than two months without drug precipitation at even higher drug loading contents than in other polymer micelles. The block copolymers are easily water-soluble and drug loading process can be performed by simple mixing in aqueous media without the use of organic solvent. The chemical composition of hydrotrope-rich core may be tailor-made to optimize and maximize the micellar property and hydrotropic activity for a specific drug. Therefore, the hydrotropic polymer micelle presents an alternative and promising approach in formulation of poorly soluble drugs.

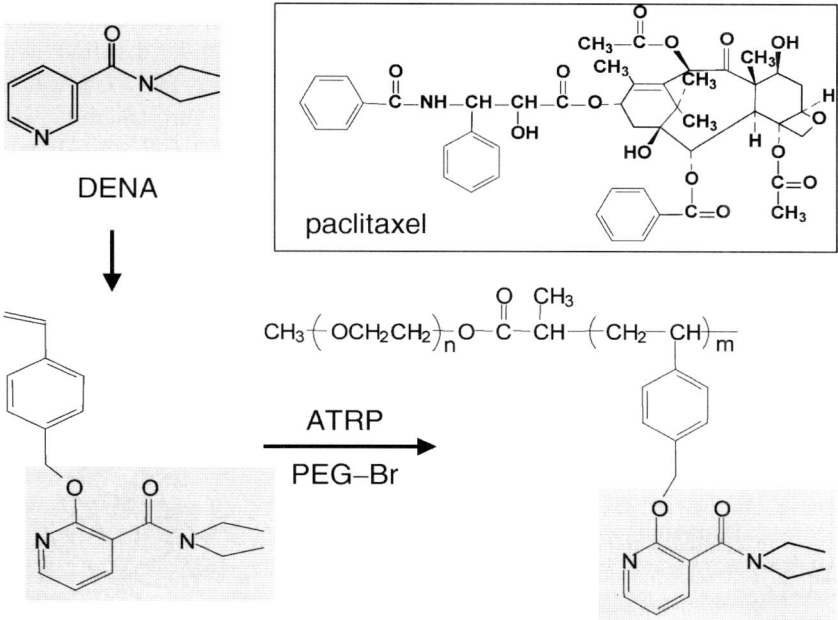

Figure 4-10. Synthetic route for amphiphilic block copolymer of poly(ethylene glycol) and hydrotropic polymer.

5.2   Hydrotropic dendrimer micelles

Instead of amphiphilic block copolymers, hydrotropic dendrimers (PGD) can be used as a building block of micelles. As mentioned in the section 4, dendritic structure of PGDs with 4-5 generations significantly increased aqueous solubility of paclitaxel. By conjugating hydrophobic moieties to PGDs, one can imagine that the hydrophobically modified PGDs form micelles, the shell part of which consists of PGDs (Fig. 4-11). As a hydrophobic moiety, cholesterol was conjugated with PGD (generation 4, G4), and its self-assembled structure was evaluated by dynamic light scattering (DLS) and atomic force microscopy (AFM) [49]. Cholesterol was selected as a hydrophobic moiety of the conjugate because cholesterol conjugated with water-soluble polymers is known to form stable aggregates in aqueous solution [50,51]. A cholesterol-conjugated PGD-G4 (Chol-PGDG4) was synthesized by conjugating cholesterol chloroformate and PGD-G4 (Fig. 4-12). To prepare self-assembled Chol-PGDG4, distilled water was added to Chol-PGD-G4 at room temperature. Chol-PGD-G4 was immediately dissolved, and the solution was stirred for 10 min. From histogram analysis of the DLS, mean diameter of Chol-PGD-G4 at the concentration of 2.6×10^{-6} M was calculated to be 49.9-59.9 nm. Although small intensity (%) of lager diameter was observed, 81% of the scattering intensity was come from the diameter of 49.9-59.9 nm. On the other hand, the diameter of PGD G-4 itself was observed at 0.99-2.3 nm. These results suggest that the conjugation of one cholesterol group to PGD-G4 induces formation of self-assembly in water. The results of AFM indicate that the mean diameter of Chol-PGD-G4 micelle was around 20 nm.

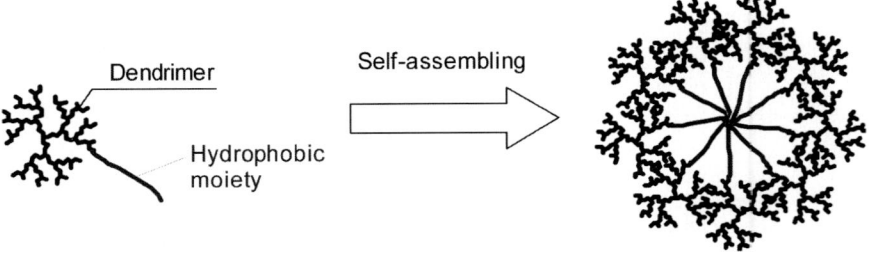

Figure 4-11. Self-assembly of hydrophobically modified PGDs to a micellar structure in aqueous solution.

From the viewpoint of the solubilizing effect of PGD-G4, association of the PGD-G4 molecules may affect the solubilizing ability. After making the Chol-PGD-G4 micelle, the solubilization test was carried out using PGD-G4 as a control. Paclitaxel solubility of Chol-PGD-G4 was almost similar to that of PGD-G4 at the concentration ranging from 0.5 to 10 wt%. This result suggests that the PGD-G4 molecules exist on outer parts of the self-assembly and act as the hydrotrope. The cholesterol group does not seem to participate in the solubilization of paclitaxel, because hydrophobic core of self-assembly can generally solubilize paclitaxel very well [52]. Presumably, once the self-assembled structure was formed, the assembly was stable in aqueous conditions due to strong hydrophobic interaction of cholesterol groups, and therefore, the outer parts consisting PGD-G4 could interact with paclitaxel.

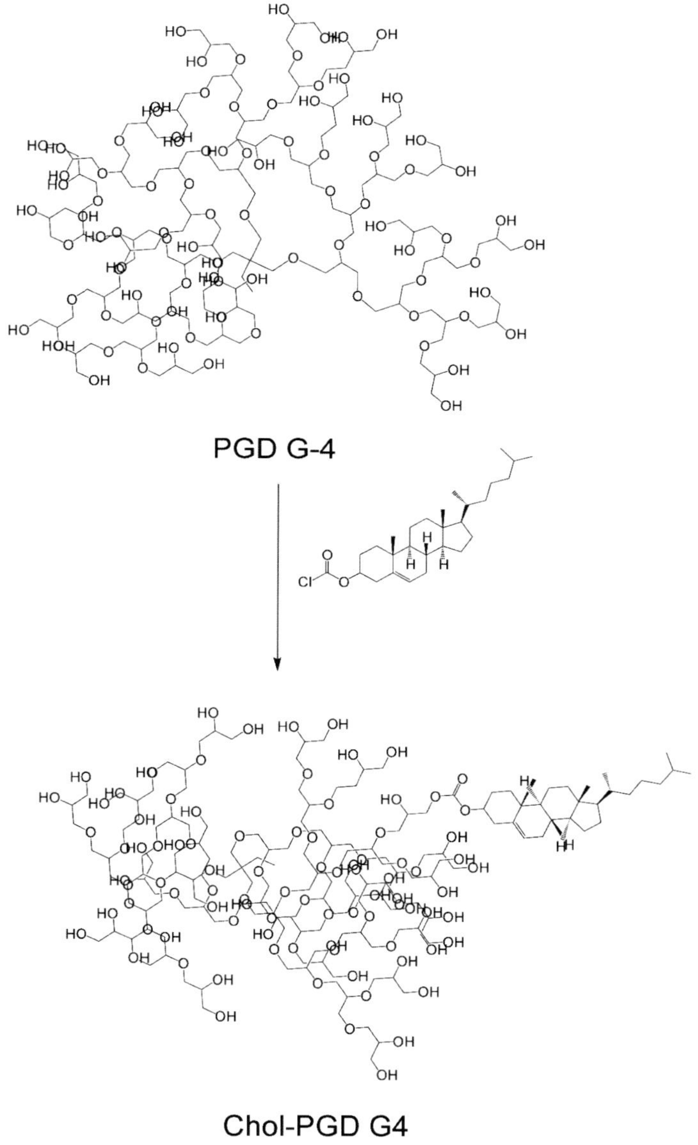

PGD G-4

Chol-PGD G4

Figure 4-12. Synthetic scheme of Chol-PGDG4. (Modified from reference 49).

6. CONCLUSIONS

Hydrotropic polymers, dendrimers and micelles have been developed as nanocarriers for poorly water soluble drugs. Paclitaxel has been used as a model hydrophobic drug to show the effectiveness of solubility enhancing properties of various hydrotropic nanocarieers. Hydrotropic polymers and dendrimers have advantages for solubilization of paclitaxel in terms of increasing the local concentration of hydrotropes in aqueous media. Those compounds were developed as hydrotropic polymer or dendrimer micelles. The hydrotropic polymer micelles present unique advantages over conventional polymer micelles in that the interaction between the polymer segment and paclitaxel is based on miscibility between the two, instead of the hydrophobic interaction alone. For this reason, the hydrotropic polymer micelles in aqueous solution are more stable than the conventional polymer micelles. The hydrotropic dendrimer micelles were formed by self-assembling cholesterol-dendrimer conjugates. The dendrimer acted as micelle shells that could solubilize paclitaxel as well as dendrimer itself. The new polymer systems based on hydrotropic polymers and dendrimers provide a new approach of designing nanocarriers for poorly soluble drugs.

Acknowledgement
This study was supported in part by National Institute of Health through grant GM65284.

7.   REFERENCES

1.   Taton, T.A. Bio-nanotechnology: Two-way traffic, *Nature Materials* 2003, *2*, 73-44.
2.   Baldwin, B.L. Making a network of hydrophobic clusters, *Science* 2002, *295*, 1657-1658.
3.   Scatena, L.F., Brown, M.G., Richmond, G.L. Water at hydrophobic surfaces: weak hydrogen bonding and strong orientation effects, *Science* 2001, *292*, 908-912.
4.   Jain, T.K., Morales, M.A., Sahoo, S.K., Lesile-Pelecky, D.L., Labhasetwar, V. Iron oxide nanoparticle for sustained delivery of anticancer drugs, *Mol. Pharm.* 2005, *2*, 185-193.
5.   Gao, Z., Lukyanov, A.N., Singhal, A., Torchilin, V.P. Diacyllipid-polymer micelles as nanocarriers for poorly soluble anticancer drugs, *Nano Lett.* 2004, *4*, 1915-1918.
6.   Srinivas, G., Discher, D.E., Klein, M.L. Self-assembly and properties of diblock copolymers by coarse-grain molecular dynamics, *Nature Materials*, 2004, *3*, 638-644.
7.   Coffman, R.E., Kildsig, D.O. Hydrotropic solubilization-mechanistic studies, *Pharm. Res.* 1996, *13*, 1460-1463.
8.   Kildsig, D.O., Suzuki, H., Sunada, H. Mechanistic studies on hydrotropic solubilization of nifedipine in nicotinamide solution. *Chem. Pharm. Bull* 1998, *46*, 125-130.

9. Silva, R.C.D., Spitzer, M., Silva, L.H.M.D., Loh, W. Investigations on the mechanism of aqueous solubility increase caused by some hydrotropes. *Thermochimica Acta* 1999, *328*, 161-167.

10. Balasubramanian, D., Srinivas, V., Gaikar, V.G., Sharma, M.M. Aggregation behavior of hydrotropic compounds in aqueous solution. *J. Phys. Chem.* 1989, *93*, 3865-3870.

11. Hussain, M.A., Diluccio, R.C., Maurin, M.B. Complexation of moricizine with nicotinamide and evaluation of the complexation constants by various methods. *J. Pharm. Sci.* 1993, *82*, 77-79.

12. Rasool, A.A., Hussain, A.A., Dittert, L.W. Solubility enhancement of some water-insoluble drugs in the presence of nicotinamide and related compounds. *J. Pharm. Sci.* 1991, *80*, 387-393.

13. Fawzi, M.B., Davision, E., Tute, M.S. Rationalization of drug complexation in aqueous solution by use of huckel frontier molecular orbitals. *J. Pharm. Sci.* 1980, *69*, 104-106.

14. Lee, J., Lee, S.C., Acharya, G., Chang, C.J., Park, K. Hydrotropic solubilization of paclitaxel: Analysis of chemical structures for hydrotropic property. *Pharm. Res.* 2003, *20*, 1022-1030.

15. Login, R.B., Merianos, J.J., Dandreaux, G., and Shih, J.S. Polymerizable derivatives of 5-oxo-pyrrolidinecarboxylic acid, U.S. Patent, U.S.A., 4,946,967, 1990.

16. Dandreaux, G., Login, R.B., Merianos, J.J., Garelick, P., Plochocka, K., Negrin, M. and Shih, J.S. Poly(pyrrolidonyl oxazoline), U.S. Patent, U.S.A., 5,008,367, 1991.

17. Lee, S.C, Lee, J., and Park, K. *Unpublished results.*

18. Lee, S.C., Acharya, G., Lee, J. and Park, K. Hydrotropic polymers: Synthesis and characterization of polymers containing picolylnicotinamide moieties, *Macromolecules* 2003, *36*, 2248-2255.

19. Frechet, J.M.J., Tomalia, D.A. *Dendrimers and Other Dendritic Polymers,* John Wiley and Sons, NY, 2001.

20. Kojima, C., Kono, K., Maruyama. K., Takagishi, T. Synthesis of polyamidoamine dendrimers having poly(ethylene glycol) grafts and their ability to encapsulate anticancer drugs, *Bioconjugate Chem.* 2000, *11*, 910-917.

21. Morgan, M.T., Carnahan, M.A., Immoos, C.E., Ribeiro, A.A., Finkelstrin, S., Lee, S.J., Grinstaff, M.W. Dendritic molecular capsules for hydrophobic compounds, *J. Am. Chem. Soc.* 2003, *125*, 15485-15489.

22. Ihre, H.R., Padilla De Jesus, O.L., Szoka, F.C. Jr., Frechet, J.M.J. Polyester dendritic systems for drug delivery applications: Design, synthesis, and characterization. *Bioconjugate Chem.* 2002, *13*, 443-452.

23. Padilla De Jesus, O.L., Ihre, H.R., Gagne, L., Frechet, J.M.J., Szoka, F.C.Jr. *Bioconjugate Chem.* 2002, *13*, 453-461.

24. Haag, R., Sunder, A., Stumbe, J.F. An approach to glycerol dendrimers and pseudo-dendritic polyglycerols. *J. Am. Chem. Soc.* 2000, *122*, 2954-2955.

25. Frey, H., Haag, R. Dendritic polyglycerol: a new versatile biocompatible material. *Rev. Mol. Biotech.* 2002, *90*, 257-267.

26. Leuner, C., Dressman, J. Improving drug solubility for oral delivery using solid dispersions, *European J. Pharm. Biopharm.* 2000, *50*, 47-60.

27. Sugimoto, M., Okagaki, T., Narisawa, S., Koida, Y., Nakajima, K. Improvement of dissolution characteristics and bioavailability of poorly water-soluble drugs by novel cogrinding method using water-soluble polymer. *Int. J. Pharm.* 1998, *160*, 11-19.

28. Basit, A.W., Newton, J.M., Short, M.D., Waddington, W.A., Ell, P.J., Lacey, L.F. The effects of polyethylene glycol 400 on gastrointestinal transit: Implications for the formulation of poorly-water soluble drugs, *Pharm. Res.*, 2001, *18*, 1146-1150.

29. Groves, M.J., Bassett, B., Sheth, V. The solubility of 17 β-oestradiol in aqueous polyethylene glycol 400, *J. Pharm. Pharmacol.*, 1984, *36*, 799-802.

30. Sato, T., Niwa, H., Chiba, A. Dynamical structure of oligo(ethylene glycol)s-water solutions studied by time domain reflectometry, *J. Chem. Phys.*, 1998, *108*, 4138-4147.

31. Ooya, T., Lee, J., Park, K. Hydrotropic dendrimers of generations 4 and 5: Synthesis, characterization, and hydrotropic solubilization of paclitaxel, *Bioconjugate Chem.*, 2004, *15*, 1221-1229.

32. Ooya, T., Lee, J., Park, K. Effects of ethylene glycol-based graft, star-shaped, and dendritic polymers on solubilization and controlled release of paclitaxel. *J. Controlled Release*, 2003, *93*, 121-127.

33. Mall, S., Buckton, G., Rawlins, D.A. Dissolution behaviour of sulphonamides into sodium dodecyl sulfate micelles: A thermodynamic approach. *J. Pharm. Sci.*, 1996. *85*, 75-78.

34. Myrdal, P.B., Yalkowsky, S.H. *Solubilization of drugs in aqueous media*, in *Encyclopedia of Pharmaceutical Technology*. 2002, Marcel Dekker, Inc. pp. 2458-2480.

35. Bader, H., Ringsdorf, H., Schmidt, B., Watersoluble Polymers in Medicine. *Angew. Makromol. Chem.*, 1984. *123*, 457-485.

36. Allen, C., Maysinger, D., Eisenberg, A. Nano-engineering block copolymer aggregates for drug delivery. *Colloids and Surfaces B: Biointerfaces*, 1999. *16*, 3-27.

37. Jones, M.C., Leroux, J.C. Polymeric micelles - a new generation of colloidal drug carriers. *Euro. J. Pharm. Biopharm.*, 1999. *48*, 101-111.

38. Lavasanifar, A., Samuel, J., Kwon, G.S. Poly(ethylene oxide)-block-poly(L-amino acid) micelles for drug delivery. *Adv. Drug. Del. Rev.*, 2002. *54*, 169-190.

39. Kwon, G.S. Polymeric micelles for delivery of poorly water-soluble compounds. *Crit. Rev. Ther. Drug Carr. Syst.*, 2003. *20*, 357-404.

40. Kwon, G.S., Okano, T. Polymeric micelles as new drug carriers. *Adv. Drug. Del. Rev.*, 1996. *21*, 107-116.

41. Soga, O., Van Nostrum, C.F., Fens, M., Rijcken, C.J.F., Schiffelers, R.M., Storm, G. and Hennink, W.E. Thermosensitive and biodegradable polymeric micelles for paclitaxel delivery. *J. Control. Rel.*, 2005. *103*, 341-353.

42. Ooya, T., Lee, J., Park, K., Solubility enhancement of paclitaxel by PEGylated polyglycerol dendrimers. *Controlled Release Society 31st Annual Meeting transactions*, 2004, #684.

43. Nakanishi, T., Fukushima, S., Okamoto, K., Suzuki, M., Matsumura, Y., Yokoyama, M., Okano, T., Sakurai, Y. and Kataoka, K. Development of the polymer micelle carrier system for doxorubicin. *J. Control. Rel.*, 2001, *74*(1-3), 295-302.

44. Yokoyama, M., Okano, T., Sakurai, Y., Suwa, S. and Kataoka, K. Introduction of cisplatin into polymeric micelle. *J. Control. Rel.*, 1996. *39*, 351-356.

45. Lavasanifar, A., Samuel, J., Kwon, G.S. Micelles self-assembled from poly(ethylene oxide)-block-poly(N-hexyl stearate L-aspartamide) by a solvent evaporation method: effect on the solubilization and haemolytic activity of amphotericin B. *J. Control. Rel.*, 2001, *77*, 155-160.

46. Ould-Ouali, L., Noppe, M., Langlois, X., Willems, B., Te Riele, P., Timmerman, P., Brewster, M.E., Arien, A., Preat, V. Self-assembling PEG-p(CL-co-TMC) copolymers for oral delivery of poorly water-soluble drugs: a case study with risperidone. *J. Control. Rel.*, 2005. *102*, 657-668.

47. Aliabadi, H.M., Mahmud, A., Sharifabadi, A.D., Lavasanifar, A. Micelles of methoxy poly(ethylene oxide)-b-poly(epsilon-caprolactone) as vehicles for the solubilization and controlled delivery of cyclosprine A. *J. Control. Rel.*, 2005. *104*, 301-311.

48. Huh, K.M., Lee, S.C., Cho, Y.W., Lee, J., Jeong, J.H., Park, K. Hydrotropic polymer micelle system for delivery of paclitaxel. *J. Control. Rel.*, 2005. *101*, 59-68.

49. Ooya, T., Huh, K.M., Saitoh, M., Tamiya, E., Park, K. Self-assembly of cholesterol-hydrotropic dendrimer conjugates into micelle-like structure: Preparation and hydrotropic solubilization of paclitaxel. *Sci. Tech. Adv. Mater.*, 2005, *6* (5): 452-456.

50. Akiyoshi, K., Deguchi, S., Tajima, H., Nishikawa, T., Sunamoto, J. Microscopic structure and thermoresponsiveness of a hydrogel nanoparticle by self-assembly of a hydrophobized polysaccharide. *Macromolecules* 1997, *30*, 857-861.

51. Yusa, S., Kamachi, M., Morishima, Y. Self-association of cholesterol-end-capped poly(sodium 2-(acrylamido)-2-methylpropanesulfonate) in aqueous solution. *Macromolecules* 2000, *33*, 1224-1231.

52. Liggins, R.T., Burt, H.M. Polyether–polyester diblock copolymers for the preparation of paclitaxel loaded polymeric micelle formulations. *Adv. Drug Delivery Review* 2002, *54*, 191-202.

Chapter 5

BIOMIMETIC APPROACH TO DRUG DELIVERY AND OPTIMIZATION OF NANOCARRIER SYSTEMS

E. Turker Baran[1,2] and Rui L. Reis[1,2]

[1]3B's Research Group-Biomaterials, Biodegradables and Biomimetics, University of Minho, Campus de Gualtar, 4710-057, Braga, Portugal; [2]Department of Polymer Engineering, Univiversity of Minho, Campus de Azurem, 4800-058, Guimaraes, Portugal

Abstract: In biomimetic approach, nanoparticulate surfaces can be covered with phospholipid bilayer(s) to obtain biological surfaces similar to cell membranes. This approach not only increases biocompatibility of drug carriers, but also provides a lipopilic medium for insertion of amphiphilic proteins which have higher affinity for target cells and channel proteins for controling solute transfer through phospholipid layer. This chapter explains recent developments in biomimetic approach to drug delivery.

Key words: Nanoparticles, Cell membranes, Phospholipid coating, Encapsulation efficiency, Surface modification.

1. INTRODUCTION

Covering the surfaces of nanocarriers with phospholipid molecules increases the stability and biocompatibility of the carrier-drug systems. Biovector biomimetic synthetic delivery system, a virus-like particle made of an inner core of polysaccharide hydrogel surrounded by a lipid bilayer, is a recent example for this new family of nanoparticulate drug carriers. It is possible to increase the encapsulation efficiency by modification of nanoparticle preparation conditions. Among these modifications are the covalent attachments of amphiphilic polymers to proteins to be encapsulated, thereby increasing the interaction between the hydrophobic polymer phase in suspension, decreasing the protein solubility by adjusting the pH of the water

75

M.R. Mozafari (ed.), Nanocarrier Technologies: Frontiers of Nanotherapy, 75–86.

phase of emulsion to the protein's isoelectric point, or by increasing polymer-drug interaction by using oppositely charged polymers. Great efforts have been made in recent years to obtain nanoparticles with reduced reticulo-endothelial uptake and prolonged circulation time. The modification of the surfaces of particles intended for intravenous injection, by a polymer, in order to provide a hydrophilic and steric barrier would minimize coating by plasma proteins, opsonization and recognition phenomena. In order to provide longer circulation time, the nanoparticles have been modified by surface coating with hydrophilic polymers (polyethylene oxide, poloxamer)/ surfactants (lauryl ethers, polysorbate) and by synthesis of biodegradable copolymers with hydrophilic segments (diblock poly(ethylene oxide)-poly(lactic acid)). Conjugation of heparin to poly (3-hydroxybutyrate-co-3-hydroxyvalerate) nanocapsules was reported to increase *in vivo* half-life and eliminating the immunogenicity of encapsulated anti-leukaemic asparaginase.

2.   APPLICATION OF CELL MEMBRANE RESEARCH TO DRUG DELIVERY

The emerging aim of controlled and targeted drug delivery is to obtain effective drug dose at the disease site and prevent side effects. Cell biology related bioevents triggered by extracellular regulators such as controlled secretion of hormones by secretory cells and a better understanding of material exchange across the cell membrane combined with the recent achievements in membrane protein science are providing important insights to researchers who try to design more biological friendly drug carriers.

Red Blood Cells (RBC) have become a useful model for pharmaceutical scientists due to possessing a surface inert to serum protein adsorption and reticuloendothelial cells. It has already been shown by several investigators that RBC can be used as carrier for drugs, such as anti-leukaemic enzyme L-asparaginase (ASNase), with a long half-life, almost as long as the native RBC half-life in human circulation [1, 2]. In addition, this surface inertness of RBC can be harnessed for prolonging intravascular particle circulation by anchoring the nanoparticles to the surface of RBCs [3].

The delivery of drugs to distant targets inside circulatory system can be possible at RBC cell membrane level as well. This kind of drug delivery system involves the transient 'loading' of RBC with a lipophilic 'hook' macromolecule which has enough affinity for the RBC plasma membrane to anchor. But given the opportunity, the macromolecule can exit its position and transfer to another (target) cell membrane for which it has a greater affinity. Binding of antimicrobial peptide, derivative of antimalarial dermaceptin S4, K-S4 (1–13)a to the plasma membrane 4 of RBC was assessed *in vitro* and *in vivo* and found to be rapid, spontaneous and receptor independent, as was the transfer of the RBC-bound peptide to the plasma membrane of microorganisms [4]. This study provided further insight to the possible use of RBC as a transport vehicle to deliver drugs to distant targets inside the body.

A possible application of membrane proteins, inserting porins (highly specific and sophisticated membrane channels that facilitate diffusion of oligosaccharides across the outer bacteria cell wall) from outer cell wall of *Escherichia coli* into the walls of liposomes, was studied [5]. Functional reconstitution of porins into the capsule shell allows control of the rate and selectivity of substrate permeation, and thus control of the enzyme reaction kinetics. Nature provides many specific, non-specific or ligand gated channel proteins, providing a unique tool to control permeation across the nano-container shells. It is now known from the human genome project that among the estimated 35000 genes there are large number of transporters responsible for movement of nutrients and absorption of many drugs [6]. Understanding the mechanism of transporters and cloning them may lead to possible use in responsive and controlled drug delivery from reconstituted carrier systems.

The control of solute transfer through mammalian phospholipid bilayers can be precisely established by parasitic microorganisms through channel proteins. Malaria-infected red blood cells (RBCs) possess parasite-induced transport pathways, which increase the uptake of selected nutrients and loss of metabolic waste products [7]. Besides, these pathways are obvious anti-malarial targets for selective inhibition of those channel proteins. For an investigator these can be invaluable routes or models for controlled and selective drug delivery using biomimetic approach.

3. PHOSPHOLIPID COATING ON POLYMERIC DRUG CARRIERS

Construction of lipid layers on biopolymers became another successful application of a biomimetic approach to drug delivery. Biovector biomimetic synthetic delivery system, called the Light Supramolecular Biovector (SMBV) is a recent example for this new family of nanoparticulate drug carriers [8, 9]. SBMV is a virus-like particle made of an inner core of polysaccharide hydrogel surrounded by a lipid bilayer and also mimicking low density lipoproteins. These nanoparticles mimic viruses, such as *influenza*, in terms of size and their supramolecular composition, with a lipid bilayer, where membrane proteins may be inserted and an internal core, where internal proteins may be loaded [10].

A multi-component drug delivery system that closely mimics the secretory granule, the lumen of which is composed of a crosslinked polyanionic condensed polymer network encapsulated within a lipid membrane, was studied by Kiser *et al.* [11]. This lipid-coated hydrogel microparticle (microgel) was triggered to release doxorubicin content by using electroporation. When the lipid-coated microgels were electroporated in a saline solution, they swelled and disrupted their bilayer coating over a period of several seconds and exchanged doxorubicin with the external plasma saline over a period of several minutes.

Endogenous transport systems such as lipoproteins have been providing insight to develop an artificial supramolecular system. Structurally, they consist of a lipid core surrounded by a monolayer of phosholipid, cholesterol and protein [12]. Doxorubicin (DOX), coupled with human low density lipoprotein (LDL), was found to be accumulated more in the liver and accumulated less in the heart than free DOX [13]. Not only LDL-DOX can exhibit similar antiproliferative effects as for DOX but in addition, reduces DOX induced cardiotoxicity [13]. Lipid nanoparticles, and lipoprotein-mimicking biovectors (LMBV) for the delivery of methotrexate were studied by Utreja *et al.* [14]. Circulation half-life of the drug was enhanced when apoprotein analogue palmitoylpolyethylene oxide was anchored and accumulated in tissues for longer periods.

Phospholipid coating on plasmid DNA adsorbed starch-chitosan nanoparticles has been investigated by our group in order to create a barrier between DNase sensitive genetic material and body fluids. This system can be expected to possess both surface properties of liposomes and drug loading effectiveness of polymeric nanoparticles. Zeta potential of starch-chitosan nanoparticles has been found to be changing from electropositive to neutral upon plasmid adsorption and this was further decreased to much more electronegative values upon phospholipid coating (Baran *et al.*, unpublished results). This system also indicated similar transfection efficiency on HeLa cells compared to uncoated plasmid-particle counterparts, indicating no inhibitory effect of phospholipids coating on genetic expression.

4.   OPTIMIZATION OF NANO ENCAPSULATION EFFICIENCY

Encapsulation of therapeutic proteins in biodegradable, biocompatible polymeric nanocapsules has been considered a safe way of delivering enzyme and protein drugs. Nanoparticles can be used as intravascular infusion preparations, injectible emulsions for both parenteral and enteric administration, as well as vaccine dosage forms for use in subcutanous or intramuscular injection [15-17]. Unprotected therapeutic proteins and polypeptides show short *in-vivo* half-life and antigenic property when they are introduced into the body. For that reason, effective delivery of protein drugs necessitates protection from the hostile immunological system of the body. Polymers of synthetic and biological origin have been used effectively in the delivery of various agents (especially drugs) to sites in the human body [18-20]. Poly(3-hydroxybutyrate-co-3-hydroxyvalerate) (PHBV) is a very promising biodegradable polymer which has been recently used as a material for the preparation of microcapsules [21], nanoparticles and microspheres [22, 23]. This polymer is entirely natural and is produced by microorganisms as well as by genetically modified plants. As an environment-friendly material it has the potential to supersede other polymers in the construction of biomedical devices and carrier systems.

Entrapment of water-soluble drugs such as proteins in nanoparticle carrier systems can be carried out through various approaches. Double emulsion (water-in-oil-in-water, w/o/w) solvent evaporation technique is more advantageous over the others because it provides higher protection for the substance to be encapsulated [24] but carries some disadvantages as well. The most important is the low entrapment efficiency (E.E.) [25, 26]. The leakage of proteins from the first water phase (w_1) to the second (w_2) during the preparation of the second emulsion is the main reason for this low entrapment efficiency.

During the process of emulsion formation it is also possible that bioactive proteins can be exposed to the denaturating effect of organic solvents and shear forces generated by mechanical mixing or ultrasonication. The ability to retain their activity after encapsulation inside the nanoparticles is definitely more important than the efficiency of protein encapsulation, because their therapeutic efficacy, and thus the biological effectiveness of the nanocapsules, depends on their activity. Even then a high encapsulation efficiency without compromising activity would be preferable because in addition to the economical aspects, a lower amount of nanocapsules, and therefore, foreign materials would be introduced to the patients.

It is possible to increase the E.E. by modification of the double emulsion preparation conditions. These modifications could be covalent attachment of amphiphilic polymers to proteins to be encapsulated, thereby increasing the interaction between the protein and the hydrophobic oil phase, or by decreasing protein solubility by adjusting the pH of the secondary water phase (w_2) to the protein's isoelectric point, a pH where proteins are the least soluble in aqueous media. Interaction between the polymer and the protein can also be significant. The study by Sendil [27] showed that the decrease in the water solubilty of tetracycline HCl achieved by neutralization with NaOH minimized the loss of drug into the aqueous phase during microcapsule formation. Song *et al.* [15] demonstrated that adjusting the pH of the aqueous solution significantly increased entrapment of (2-(4-morpholinyl)-8-(3-pyridinylmethoxy)-4H-1-benzopyran-4-one) in PLGA nanoparticles (prepared by the oil-in-water technique). The solubility of the drug drops rapidly as the pH goes from neutral to basic.

Gaspar *et al.* [28] found that the presence of free carboxyl end groups in the PLGA chain affected the encapsulation efficiency, the highest encapsulation efficiency being obtained with the polymer having the most carboxyl groups. When the carboxyl groups were esterified the enzyme loading in PLGA nanocapsules was reported to decrease. It is also possible to increase E.E. considerably by the modification of charges on protein surface leading to an increase in the electrostatic interaction between the protein and the polymer's polar group. Calvo *et al.* [29] reported that E.E. for bovine serum albumin (BSA) entrapped in chitosan-polyethylene oxide nanoparticles increased when the pH was higher than the isoelectric point of the protein. At pH values higher than the isoelectric point, BSA becomes negatively charged and as a result of increased electrostatic interaction between BSA and chitosan's amine groups higher encapsulation efficiency can be obtained.

A detailed study has been carried out towards improving the encapsulation efficiency of BSA and some model enzymes (i.e. catalase, L-asparaginase and glucose oxidase) in PHBV nanocapsules as a preliminary step towards nanocapsular cancer therapy [30]. In order to improve the activity and encapsulation efficiency of these bioactive agents some parameters such as pH, relative proportions of the ingredients and the molecular weight of the polymer were systematically changed. Chemical modification of enzyme properties by conjugation with polyethylene was also tested.

For the preparation of low molecular weight PHBV, the polymer was treated with sodium borohydride. A 14-fold decrease in molecular weight (from 297,000 to 21,000) was observed upon 4h of reaction time. Although the amount of encapsulated protein was not increased, the enzyme activity was increased upon use of low molecular weight PHBV indicating that these nanocapsules have a higher permeability to solutes (reactants and products). The adjustment of the w_2 phase to isoelectric point of the proteins significantly increased the encapsulation yields of catalase, L-asparaginase and BSA. Likewise, polyethylene glycol coupling significantly increased the entrapment efficiency as well as the activity of catalase and L-asparaginase. A combination of the various optimum preparation conditions further increased the encapsulated catalase activity (about six-fold) in comparison to the initial conditions (with no modification and no isoelectric point adjustment).

5. SURFACE MODIFICATION OF DRUG CARRIERS WITH HYDROPHILIC POLYMERS

Only about one part in 10,000 of an intravenously injected agent reaches its final cellular target when that target is located at a deep tissue site [31]. Up to 95% of loss occurs as a result of the initial encounter of the injected agent with plasma. Specific bio-distribution and site targeting are enabled only when an agent or delivery system resists opsonisation, and precipitation by plasma proteins and the aqueous milieu of plasma. After the particles are injected into the blood stream they are rapidly conditioned (or coated) by plasma proteins and glycoproteins. This process, known as opsonisation, is critical in dictating the subsequent fate of the administered particles into circulation. Normally for nanoparticles and other foreign materials, that are recognized as foreign bodies, the opsonisation process makes the injected material easily recognizable for the major defense system of the body – i.e. the reticuloendothelial system (RES) or the mononuclear phagocyte system. Great efforts have been made in recent years to obtain biodegradable nanospheres and microspheres with reduced RES uptake and prolonged circulation time [32–35]. In addition, drug targeting usually requires vectors having long enough circulation in the body to reach their site of action. For all these reasons, the development of carrier systems disclosing a surface that does not adsorb serum proteins and is non-antigenic is a crucial requirement for long half-life in a hostile body environment.

The modification of the surfaces of the particles intended for intravenous injection, by a polymer, in order to provide a hydrophilic and a steric barrier would minimize coating by the plasma components, opsonisation and recognition phenomena. This concept of covering particles with a hydrophilic surface coating has been applied to biodegradable particles as well as to liposomes and emulsions [36]. It was possible to cover particles by making their surfaces both hydrophilic and sterically stable using polymer adsorption technique. Particles that carry adsorbed or grafted layers of polymers can display longer stability characteristics because of the unfavorable thermodynamics of opsonins approaching the adsorbed or grafted polymeric chains.

Surface modification of nanoparticles by biodegradable and long circulating polymers has been achieved mainly by two methods: (i) surface coating with hydrophilic polymers/surfactants and (ii) development of biodegradable copolymers with hydrophilic segments. Some of the widely used surface-coating materials are: poly(ethylene oxide) (PEO), poly(ethylene glycol) (PEG) [34, 35–39], poloxamer–polyamine [38-41], lauryl ethers (Brij-35) and polysorbate (Tween-80) [42, 43]. Polaxamers and polyamines adsorb strongly onto the surface of hydrophobic nanospheres e.g. polystyrene, poly(lactide-co-glycolide), poly(phosphazene), poly(methyl methacrylate) and poly(butyl 2-cyanoacrylate) nanospheres – via their hydrophobic Poly(oxypropylene) (POP) centre block. This kind of adsorption leaves the hydrophilic PEO side-arms in a mobile state because they extend outwards from the particle surface to aqueous surrounding [41].

Nanoparticles covered with poly(ethylene oxide) (PEO) or PEG chains on their surface have been described [44-47] as blood persistent drug delivery systems with potential applications for intravenous drug administration. Gref et al. [34] prepared diblock poly(ethylene oxide)-poly(lactic acid), poly(ethylene oxide)-poly(lactide-co-glycolide) and poly(ethylene oxide)-poly(caprolactone) nanoparticles to investigate affect of PEO corona thickness on protein adsorption. It was reported that increasing the PEO content in the nanoparticles above 5% did not cause any further reduction in protein adsorption. Intravenously injected doxorubicin-loaded polysorbate 80-coated nanoparticles were able to lead to a 40% cure in rats with intracranially transplanted glioblastomas [43]. Over coating with this surfactant seems to lead to the adsorption of lipoprotein E from plasma onto the nanoparticle surface. The particles then seem to mimic low-density lipoprotein particles (LDL) and could interact with LDL receptor leading to uptake by endothelial cells of brain.

Peptides or proteins are attracting more attention as drugs since their functions are better understood and progress is made in the fields of biotechnology and bioengineering. Particularly, the development of recombinant DNA technology has made these compounds available on a larger scale compared to the past. However, unprotected therapeutic polypeptides possess a short half-life and exert antigenic property when introduced into the body. For that reason, effective administration of protein drugs necessitates their protection from the hostile immunological system of the body.

The encapsulation of L-asparaginase (ASNase) (an antitumor agent effective in human acute leukemia) into poly(3-hydroxybutyrate-co-3-hydroxyvalerate) (PHBV) nanocapsules was studied by Baran et al., in order to diminish immunogenic, toxic property of the enzyme and increase its circulation half-life [30]. The heterologous origin of the enzyme leads to an immune response upon its application to the body. Anaphylactic reactions may be observed in up to 33% of patients because of its bacterial origin that can stimulate production of immunoglobulin E (IgE) or other immuno-globulins.

Surface modification of nanocapsules was performed to increase the hydrophilicity and prevent easy opsonisation by the reticuloendothelial system. For that purpose, low molecular weight PHBV was prepared and conjugated to low molecular weight heparin. In vivo studies were performed to compare and test encapsulated L-asparaginase preparations in mice model. These were compared in terms of immunogenicity of the preparations and their pharmacodynamic properties in circulation. The enzyme activity in the blood due to unmodified PHBV nanocapsules dropped to 38% of its initial value 4 hours after injection. When the same sample was tested for the enzyme content in the circulation by using the radiolabelled enzyme a much lower enzyme (30% of initial) was shown after 3 hours. PHBV nanocapsules with heparin conjugated on their surface had a longer presence in the circulation than unmodified PHBV nanocapsules. Most importantly, no adverse effects and symptoms of anaphylaxis were observed upon injection of encapsulated ASNase-PHBV nanocapsules to mice intravenously through the tail vein. This study proved that nano-encapsulation can prevent lethal anaphylaxis responses by isolating the protein drug from the immune system elements with the help of a polymeric shell.

6.   REFERENCES

1. Kravtzoff R, Colombat PH, Desbois I, Linasser C, Muh JP, Philip T, Blay JY, Gardenbas M, Poumier-Gaschard P, Lamagnere JP, Chassaigne M, Ropars C. Tolerance evaluation of L-asparaginase loaded in red blood cells. *Eur J Pharm* 1996; 51: 221–5.

2. Satterfield W, Keeling M. Intraperitoneal administration of carrier erythrocytes in dogs: an improved method for delivery of L-asparaginase 2. *Appl Biochem* 1990; 12: 331–5.

3. Chambers E, Mitragotri S. Prolonged circulation of large polymeric nanoparticles by non-covalent adsorption on erythrocytes *J Control Rel* 2004; 100: 111–19.

4. Feder R, Nehushtai R, Mor A. Affinity driven molecular transfer from erythrocyte membrane to target cells. *Peptides* 2001; 22: 1683–90.

5. Winterhalter M, Hilty C, Bezrukov SM, Nardin C, Meier W, Fournier D. Controlling membrane permeability with bacterial porins: application to encapsulated enzymes. *Talanta* 2001; 55: 965–71.

6. Lee VHL, Sporty JL, Fandy TE. Pharmacogenomics of drug transporters: the next drug delivery challenge. *Adv Drug Del Rev* 2001; 50: S33–40.

7. Thomas SLY, Egee S, Lapaix F, Kaestner L, Staines HM, Ellory JC. Malaria parasite plasmodium gallinaceum up-regulates host red blood cell channels. *FEBS Lett* 2001; 500: 45–51.

8. Nagaich S, Khopade AJ, Jain NK. Lipid grafts of egg-box complex: a new supramolecular biovector for 5-fluorouracil delivery. *Pharm Acta Helv* 1999; 73: 227–36.

9. Major M, Prieur E, Tocanne JF, Betbeder D, Sautereau AM. Characterization and phase behaviour of phospholipid bilayers adsorbed on spherical polysaccharidic nanoparticles. *Biochim Biophys Acta* 1997; 1327: 32–40.

10. Hoegen PV. Synthetic biomimetic supramolecular biovector (SMBV) particles for nasal vaccine delivery. *Adv Drug Del Rev* 2001; 51: 113–25.

11. Kiser PF, Wilson G, Needham D. Lipid-coated microgels for the triggered release of doxorubicin. *J Control Rel* 2000; 68: 9–22.

12. Rensen PCN, de Vrueth RLA, Kuiper J, Bijsterbosch MK, Biessen EAL, van Berkel TJC. Recombinant lipoproteins: lipoprotein-like lipid particles for drug targeting. *Adv Drug Del Rev* 2001; 47: 251–76.

13. Chu ACY, Tsabg SY, Lo EHK, Funk KP. Low density lipoprotein as a targeted carrier for doxorubicin in nude mice bearing human hepatoma HepG2 cells. *Life Sci* 2001; 70: 591–601.

14. Utreja S, Khopade AJ, Jain NK. Lipoprotein-mimicking biovectorized systems for methotrexate delivery. *Pharma Acta Helvetiae* 1999; 73: 275–79.

15. Song C, Labhasetwar V, Murphy H, Qu X, Humphrey WR, Shebuski RJ, Levi RJ. Formulations and characterisation of biodegradable nanoparticles for intravascular local drug delivery. *J Control Rel* 1997; 43: 197–212.

16. Gibaud S, Rousseau C, Weingarter C, Favier R, Douay L, Andreux JP, Couvreur P. Polyalkylcyanoacrylate nanoparticles as carriers for granulocyte-colony stimulating factor (G-CSF). *J Control Rel* 1998; 52: 131–39.

17. Lemoine D, Preat V. Poymeric nanoparticles as the delivery system for influenza virus gycoproteins. *J Control Rel* 1998; 54: 15–27.

18. Lin W, Garnett MC, Davies MC, Bignotti F, Fernutti P, Davis SS, Illum L. Preparation of surface modified albumin nanospheres. *Biomaterials* 1997; 18: 559–65.

19. Jeong Y, Cheon J, Kim S, Nah J, Lee Y, Sung Y, Akaike T, Cho C. Clonazepam release from core-shell type nanoparticles *in vitro*. *J Control Rel* 1998; 51: 169–78.

20. Aboubakar M, Puisieux F, Couvreur P, and Vauthier C. Physico-chemical characterization of insulin-loaded poly(isobutylcyanoacrylates) nanocapsules obtained by interfacial polymerization. *Int J Pharm* 1999; 183: 63–66.

21. Sendil D, Gursel I, Wise DL, Hasirci V. Antibiotic release from biodegradable PHBV microparticles. *J Control Rel* 1999; 59: 207–17.

22. Pouton CW, Akhtar S. Biosynthetic polyhydroxyalkonates and their potential in drug delivery. *Adv Drug Deliver Rev* 1996; 18: 133–62.

23. Gangrade N, Price JC. Poly(hydroxybutyrate-hydroxyvalerate) microspheres containing progesterone: preparation, morphology and release properties. *J Microencapsul* 1991; 8: 185–202.

24. Embleton JK, Tighe BJ. Polymers for biodegradable medical devices. Encapsulation studies: control of poly-hydroxybutyrate-hydroxyvalerate microcapsules porosity via polycaprolactone blending. *J Microencapsul* 1993; 10: 341–52.

25. Conway BR, Eyles JE, Alpar HO. A comparative study on the immune responses to antigens in PLA and PHB microspheres. *J Control Rel* 1997; 49: 1–9.

26. Atkins TW. Fabrication of microcapsules using poly(ethylene adipate) and a blend of poly(ethylene adipate) with poly(hydroxybutyrate-hydroxyvalerate). Incorporation and release of bovine serum albumin. *Biomaterials* 1997; 18: 173–80.

27. Sendil D. 1997; *Antibiotic release from biodegradable microbial polyesters*. M.S. thesis, Middle East Technical University, Ankara, Turkey.

28. Gaspar MM, Blanco D, Cruz ME, Alonso MJ. Formulation of L-asparaginase-loaded poly(lactide-co-glycolide) nanoparticles: influence of polymer properties on enzyme loading, activity and *in vitro* release. *J Control Rel* 1998; 52: 53–62.

29. Calvo P, Remunan-Lopez, C, Vila-Jato JL, Alonso MJ. Novel hydrophilic chitosan-polyethylene oxide nanoparticles as protein carriers. *J Appl Polym Sci* 1997; 63: 125–32.

30. Baran ET, Ozer N, Hasirci V. Nanoencapsulation of asparaginase into Poly(hydroxy-butyrate-hydroxy-valerate) nanocapsules. *J Mater Sci: Mater Med* 2002; 12: 1113–21.

31. Ranney DF. Biomimetic transport and rational drug delivery. *Biochem Pharma* 2000; 59: 105–14.

32. Kim ES, Lu C, Khuri FR, Tonda M, Glisson BS, Liu D, Jung M, Hong WK. A phase II study of STEALTH cisplatin (SPI-77) in patients with advanced non-small cell lung cancer. *Lung Cancer* 2001; 34: 427–32.

33. Yamamoto Y, Nagasaki Y, Kato Y, Sugiyama Y, Kataoka K. Long circulating poly(ethylene glycol)-poly(DL-lactide) block copolymers micelles with modulated surface charge. *J Control Rel* 2001; 77: 27–38.

34. Gref R, Luck M, Quellec P, Marchand M, Dellacherie E, Harnisch S, Blunk T, Muller RH. Stealth' corona-core nanoparticles surface polymodified by polyethylene glycol (PEG): influences of the corona (PEG chain length and surface density) and of the core composition on phagocytic uptake and plasma protein adsorption. *Colloid Surf B Biointerf* 2000; 18: 301–13.

35. Li YP, Pei YY, Zhang XY, Gu ZH, Zhou ZH, Yuan WF, Zhou JJ, Zhu JH, Gao XJ. PEGYlated PLGA nanoparticles as protein carriers: synthesis, preparation and biodistribution in rats. *J Control Rel* 2001; 71: 203–11.

36. Kataoka K, Harada A, Nagasaki Y. Block copolymer micelles for drug delivery: design, characterization and biological significance. *Adv Drug Del Rev* 2001; 47: 113–31.

37. Calvo P, Gouritin B, Brigger I, Lasmezas C, Deslys J, Williams A, Andreux JP, Dormont D, Couvreur P. PEGy lated polycyanoacrylate nanoparticles as vector for drug delivery in prion diseases. *J Neurosci Meth* 2001; 111: 151–5.

38. Brigger I, Chaminade P, Marsaud V, Appel M, Besnard M, Gurny R, Renoir M, Couvreur P. Tamoxifen encapsulation within polyethylene glycol-coated nanospheres: a new antiestrogen formulation. *Int J Pharm* 2001; 214: 37–42.

39. Mosqueira VCF, Legrand P, Gulik A, Bourdon O, Gref R, Labarre D, Barratt G. Relationship between complement activation, cellular uptake and surface physico-chemical aspects of novel PEG-modified nanocapsules. *Biomaterials* 2001; 22: 2967–79.

40. Redhead HM, Davis SS, Illum L. Drug delivery in poly(lactide-coglycolide) nanoparticles surface modified with poloxamer 407 and poloxamine 908: *in vitro* characterization and in vivo evaluation. *J Control Rel* 2001; 70: 353–63.

41. Stolnik S, Daudali B, Arien A, Whetstone J, Heald CRM, Garnett C, Davis SS, Illum L. The effect of surface coverage and conformation of poly(ethylene oxide) (PEO) chains of

poloxamer 407 on the biological fate of model colloidal drug carriers. *BBA Biomembranes* 2001; 1514: 261–79.

42. Kreuter J, Petrov VE, Kharkevich DA, Alyautdin RN. Influence of the type of surfactant on the analgesic effects induced by the peptide dalargin after its delivery across the blood–brain barrier using surfactant-coated nanoparticles. *J Control Rel* 1997; 49: 81–7.

43. Gelperina SE, Khalansky AS, Skidan IN, Smirnova ZS, Bobruskin AI, Severin SE, Turowski B, Zanella FE, Kreuter J. Toxicological studies of doxorubicin bound to polysorbate 80-coated poly(butylcyanoacrylate) nanoparticles in healthy rats and rats with intracranial glioblastoma. *Toxicol Lett* 2002; 126: 131–41.

44. Photos PJ, Bacakova L, Discher B, Bates FS, Discher DE. Polymer vesicles *in vivo*: correlations with PEG molecular weight. *J Control Rel* 2003; 90: 323–34.

45. Moghimi SM, Hunter AC. Poloxamers and polyamines in nanoparticle engineering and experimental medicine. *TIBTECH* 2000; 18: 412–20.

46. Zambaux MF, Bonneaux F, Gref R, Dellacherie E, Vigneron C. Protein C-loaded monomethoxypoly (ethylene oxide)–poly(lactic acid) nanoparticles. *J Pharm* 2001; 212: 1–9.

47. Maiti S, Jayachandran KN, Chatterji PR. Probing the association behaviour of poly(ethyleneglycol) based amphiphilic comb-like polymer. *Polymer* 2001; 42: 7801–8.

Chapter 6

A ROLE FOR PREBIOTICS IN CONTROLLED DRUG DELIVERY

Ajay Awati and Paul J. Moughan
Riddet Centre, Massey University, Private bag 11222, Palmerston North, New Zealand

Abstract: Prebiotics offer several strategic benefits in terms of the intestinal health and balance in microbial community. Prebiotics as nutriceuticals have been widely investigated over last couple of decades. However, their probable use in site-specific delivery remains obscured. This chapter discusses the prebiotics, their nutritional benefits and possibilities for a role in site-specific drug delivery.

Keywords: Prebiotics, encapsulation, drug-delivery.

1.   INTRODUCTION

Since the first development of medicinal systems in different civilizations, till date, oral delivery has always been the preferred route of drug administration. Oral administration leads to greater convenience, potentially less pain to the patient, and a reduced risk of cross infection, mostly associated with parenteral drug administration [1, 2]. Despite these advantages, physiological barriers such as the stomach environment and gastrointestinal enzyme activity are not always favourable for drug administration at a specific site. This has led to an extensive amount of research in the field of site-specific drug delivery with special attention being given to micro-encapsulation of the active drug compound. In previous chapters, various authors have discussed the possibilities of using different compounds for drug encapsulation. This chapter deals with "Prebiotics" as a group of quite diverse compounds but having common properties. Prebiotics, although having been studied over the past two decades and although offering contain strategic benefits, have not been widely investigated for a role in site-specific drug delivery.

M.R. Mozafari (ed.), Nanocarrier Technologies: Frontiers of Nanotherapy, 87–94.

2.   PREBIOTICS

Gibson and Roberfroid [3] have defined the term prebiotic as: 'a non digestible food ingredient that beneficially affects the host by selectively stimulating the growth and/or activity of one or a limited number of bacteria in the colon that can improve host health.' Emphasis on the colon may be too restrictive as many monogastric species, and sometimes humans support a considerable degree of bacterial fermentation in the upper digestive tract [4, 5]. Additionally, in recent times focus has shifted towards the concept of the need for 'a stable and complex commensal bacterial community' as a pre-requisite for a healthy gut eco-system [6, 7]. Such considerations have led to further elaboration of the definition of a prebiotic [8] 'a non digestible dietary ingredient that beneficially affects the host by stimulating the activity, in terms of fermentation end products, and stability of the diverse comensal microbiota in different parts of the gastro-intestinal tract, depending on the fermentability of the dietary ingredient itself.'

3.   PREBIOTICS IN NUTRITION

The sole purpose of a prebiotic from a physiological point of view is to provide a preferred substrate for the potentially beneficial microbiota. Most of the compounds investigated for such prebiotic properties, have been dietary fibres. Soluble fibres are in general better energy substrates for gastrointestinal micro-organisms than are insoluble fibres [9]. Different soluble fibre components and their sources are shown in Table 6-1. As reviewed by Cummings *et al.* [10] ingestion of prebiotics is followed by an increased excretion of breath-hydrogen. This is evidence that these prebiotic fibre components are generally well fermented in the gastrointestinal tract, but the rate of fermentation may vary depending on the source of the fibre component, and the host species.

The important role that prebiotics play as a functional food ingredient and their health benefits in general are well documented [11-20]. Exploitation of prebiotics as a preferred substrate by commensal microbiota for site-specific drug delivery, however, is an area ripe for further investigation.

4.   PREBIOTICS FOR ENCAPSULATION

As reviewed by Fahmy *et al.* [21], one of the current challenges in drug delivery systems is finding a means of carrying the drug in a stable form to specific sites. It is important that the particular means of carriage protect the drug from immunological as well as physiological processes such as digestion, until it reaches the desired site. Using prebiotics for encapsulation is based on the simple principle, that prebiotic material can be used to coat a drug or bioactive compound, thus allowing the active compound to pass through the stomach and upper digestive tract without release. Prebiotics which have the property that they are indigestible substrates (resistant to mammalian gut enzymes) can easily pass the gastric barriers and be fermented by microbiota to release the drug. In retrospect, over the last decade several authors have reported on a sporadic basis, the use of dietary fibre components for encapsulation. Successful outcomes have been described. The idea of using microbiota/dietary interaction especially for colonic drug delivery in humans has been reviewed by Sinha and Kumaria [22]. The authors discuss three different systems that have been developed and studied by several researchers for colon specific drug delivery based on exploiting the microbial population residing in the gastro intestinal tract.

Table 6-1. Different fibre components that are well fermented by the mammalian GIT microbiota and their dietary source.

Fibre Component	Source
Gums	leguminous seed plants (guar, locust bean), seaweed extracts (carrageenan, alginates), plant extracts (gum acacia, gum karaya, gum tragacanth), microbial gums (xanthan, gellan)
β-glucans	grains (wheat, barley, oat, rye)
Pectins	fruits, vegetables, legumes, sugar beet, potato
Inulin	chicory, Jerusalem artichoke, onions, wheat
Oligosaccharides/analogues	various plants and synthetically produced
Xylo oligosaccharides	cereals
Lactulose	lactose
Isomalto oligosaccharides	algae
Gluco oligosaccharides	sucrose
Fructo oligosaccharides	sucrose
Chondroitin	animal origin

i) *Prodrugs*: A prodrug is a pharmacologically inactive derivative of a parent drug molecule that requires spontaneous or enzymatic transformation *in vivo* to release the active drug. For the colonic delivery of drugs, prodrugs are designed to undergo minimal absorption and hydrolysis in the upper GIT but to undergo enzymatic hydrolysis in the colon, thereby releasing the active drug moiety from the carrier. Inspite of high site specificity, Sinha and Kumaria [22] admonish that the prodrug approach is not very versatile as the ability to form a prodrug depends upon the functional groups available on the drug moiety for chemical linkage. Furthermore, prodrugs are new chemical compounds and thus need extensive evaluation before being used as drug carriers.

ii) *Azo polymeric prodrugs/azo polymeric coating*: Both synthetic as well as naturally occurring polymers are used as drug carriers for drug delivery to the colon. Synthetic polymers have been used to form polymeric prodrugs with azo-linkages between the polymer and the drug moiety. The metabolism of azo compounds by intestinal bacteria has been studied extensively and numerous polymers have been evaluated for the purpose. These polymers have similar shortcoming as prodrugs. Polymeric prodrugs being new chemical entities require a detailed toxicological evaluation before being used as drug delivery systems.

iii) *Polysaccharide based delivery systems*: Use of naturally occurring polysaccharides is a promising approach for future colon specific drug delivery. These compounds are inexpensive, found in abundance, and are available in a variety of structures and with varied properties [23]. Many of the polysaccharide-based delivery systems are resistant to digestion in the upper GIT and when they arrive in the colon the glycosidic linkages are hydrolysed by colonic microbiota to release the drug. Several different polysaccharides that have been used in drug delivery systems, are reviewed by Sinha and Kumaria [22]. These polysaccharides mainly include naturally occurring polysaccharides obtained from plants (guar gum, inulin), animals (chitosan, chondroitin sulphate), algae (alginates) or microbes (dextran).

Some of the naturally occurring as well as synthetic polysaccharides have been studied extensively for their prebiotic properties in humans as well as in animals. These include inulin, guar gum, resistant starch, pectins and lactulose.

In recent studies, it has been shown that the drug release from amorphous amylose coated products was accelerated in the fermentative environment of the colon. This was attributed to the bacterial digestion of the amylose component of the film coat producing pores for drug diffusion [24].

In humans, and considering the lower amount of microbial activity in the small intestine, application of this approach to drug delivery is mainly restricted to the large intestine. However, in specific disease conditions such as small bowel bacterial overgrowth, in which the main consequence of small bowel dysmotility is overgrowth of the microbial community, microbial activity in the small intestine increases [25]. In such a situation, oral dosing of an antimicrobial drug coated with fast fermenting indigestible soluble fiber, may improve the efficacy of the antimicrobial drug by site specific delivery. This is a 'Trojan horse' approach to combating pathogens. The bacteria degrade the prebiotic carrier system containing the anti-microbial drug, which is akin to delivering a Trojan horse leading to their self destruction.

There are several other strategies for site-specific drug delivery among which liposomes [26] and liposome-based carrier systems (e.g. archaeo-somes, virosomes, transferosomes) [27] are the most studied and widely applied technologies. As discussed by Mozafari *et al.* [27], these kinds of systems may have problems due to toxicities related to product deve-lopment procedures. Moreover, due to low pH and the presence of digestive enzymes (e.g. lipases) these lipidic carriers cannot generally pass intact to the lower gastrointestinal tract. In relation to these issues, prebiotics being natural dietary ingredients hold a distinct advantage both in terms of safety and non degradability.

Furthermore in veterinary therapeutics using fermentable substrates with variable fermentabilities as coatings will be an advantage. Different animal species have varied active microbial communities in different parts of the gastrointestinal tract other than the colon. For instance, chickens have a very well developed microbial community in the crop, and in the caecum, whereas the pig has active microbial fermentation in the stomach, ileum and the large intestine.

5.   FERMENTABILITY OF PREBIOTICS

Using the *in vitro* gas production technique, Awati *et al.* [28] have shown that different fermentable carbohydrates are fermented in different parts of the GIT of weanling piglets. Detailed knowledge on the fermentability of different prebiotic substances, which can be suitable candidates for encapsulation, is necessary. Knowing the probable site of degradation of the particular candidate substance would add to the accuracy of prediction of drug delivery by this approach.

Several *in-vitro* techniques for determining rate and degree of fermentability have been suggested. Coles *et al.* [29] have summarized some of the most common techniques used for the study of fermentability of different dietary fermentable fibres. Among the different techniques, the *in vitro* gas production technique has been recently used to demonstrate differences in the fermentability of different carbohydrate substrates [30, 31].

The use of *in vitro* fermentation techniques coupled with obtaining inocula from specific sites in the GIT to measure fermentability of candidate substrates and thus be able to predict the probable site of fermentation in the GIT, will lead to more accurate site-specific drug delivery in the GIT.

6.   CONCLUDING REMARKS

Prebiotic encapsulation holds a future for the delivery of bioactive materials. This approach has dual benefits in terms of a general health effect of prebiotic consumption and the prebiotic being a preferred substrate for microbiota, ensuring safe targeted drug delivery. Exploitation of microbial activity in terms of site-specific delivery, however, will need to rely on an extensive knowledge of the fermentability of the candidate substrate and the species composition and activity of the microbiota at specific sites of the gastrointestinal tract. These characteristics in turn will vary with the host animal species.

7.   REFERENCES

1.   Chen, H. & R. Langer, (1998) Oral particulate delivery: status and future trends. *Advanced Drug Delivery Reviews*, 34(2-3): 339-350.
2.   Florence, A.T. & P.U. Jani, (1993) Particulate delivery: the challenge of the oral route, in *Pharmaceutical particulate carriers: therapeutic applications*, A. Rolland, Editor, Marcel Dekker: New York: 65-107.
3.   Gibson, G.R. & M.B. Roberfroid, (1995) Dietary modulation of the human colonic microbiota: Introducing the concept of prebiotics. *Journal of Nutrition*, 125: 1401-1412.
4.   Wilson, K.H., (1997) Biota of human gastrointestinal tract, in *Gastrointestinal Microbiology*, R.I. Mackie, B.A. White, & R.E. Isaacson, Editors, Chapman and Hall: New York: 39-58.
5.   Gaskins, H.R., (2001) Intestinal bacteria and their influence on swine growth, in *Swine Nutrition*, A.J. Lewis & L.L. Southern, Editors, CRC Press LLC, Florida.: 585-608.
6.   Konstantinov, S.R., W.M. De Vos, A.D.L. Akkermans, H. Smidt, C.F. Favier, W.Y. Zhu, B.A. Williams, J. Kluss, & W.-B. Souffrant, (2004) Microbial diversity studies of the porcine gastrointestinal ecosystem during weaning transition. *Animal Research*: 317-324.
7.   Verstegen, M.W.A. & B.A. Williams, (2002) Alternatives to the use of antibiotics as growth promoters for monogastric animals. *Animal Biotechnology*: 113-127.
8.   Awati, A., (2005) Prebiotics in piglet nutrition? fermentation kinetics along the GI tract, in *PhD thesis Animal nutrition group,* Wageningen University: Wageningen, The Netherlands.
9.   Fahey, J.G.C., E.A. Flickinger, C.M. Grieshop, & K.S. Swanson, (2004) The role of dietary fibre in companion animal nutrition, in *Dietary Fibre: bio-active carbohydrates*

for food and feed, J.W. Kamp van der, N.G. Asp, J. Miller Jones, & G. Schaafsma, Editors, Wageningen Academic Publisher: Wageningen, The Netherlands.: 295-328.

10. Cummings, J.H., G.T. Macfarlane, & H.N. (2001) Englyst, Prebiotic digestion and fermentation. *American journal of clinical nutrition.* 73: 415S-420S.

11. Andrieux, C. (2001) Prebiotics and health. in *Post-Antibiotics era of animal nutrition: The 3rd Techno World Meeting.* Korea: CTC-BIO.

12. Bengmark, S. (2002) Gut microbial ecology in critical illness: Is there a role for prebiotics, probiotics, and synbiotics? *Current Opinion in Critical Care.* 8: 145-151.

13. Berg, R.D. (1998) Robiotics, prebiotics or 'conbiotics'? *Trends in Microbiology.* 6(3): 89-92.

14. Blaut, M. (2002) Relationship of prebiotics and food to intestinal microflora. *European Journal Nutrition.* 41(Suppl 1): I11-I16.

15. Chow, J. (2002) Probiotics and prebiotics: A brief overview. *Journal of Renal Nutrition.* 12(2): 76-86.

16. Collins, M.D. & G.R. Gibson (1999) Probiotics, prebiotics and synbiotics: approaches for modulating the microbial ecology of the gut. *American journal of clinical nutrition.* 69 (suppl): 1052S-1057S.

17. Fooks, L.J., R. Fuller, & G.R. Gibson (1999) Prebiotics, probiotics and human gut microbiology. *International dairy journal.* 9: 53-61.

18. Macfarlane, G.T. & J.H. Cummings (1999) Probiotics and prebiotics: can regulating the activities of intestinal bacteria benefit health? *British Medical Journal.* 318: 999-1003.

19. Rastall, R.A. & V. Maitin (2002) Prebiotics and synbiotics: towards the next generation. *Current Opinion in Biotechnology.* 13(5): 490-496.

20. Roberfroid, M. (2002) Functional food concept and its application to prebiotics. *Digestive and Liver Disease.* 34 Suppl 2: S105-110.

21. Fahmy, T.M., P.M. Fong, A. Goyal, & W.M. Saltzman, (2005) Targeted for drug delivery. *Materials Today.* 8(Suppl. 1): 18-26.

22. Sinha, V.R. & R. Kumria (2003) Microbially triggered drug delivery to the colon. *European Journal of Pharmaceutical Sciences.* 18(1): pp. 3-18.

23. Hovgaard, L. & H. Brondsted (1996) Current applications of polysaccharides in colon targeting. *Critical Reviews in Therapeutic Drug Carrier Systems.* 13: 185-223.

24. Wilson, P.J. & A.W. Basit, (2005) Exploiting gastrointestinal bacteria to target drugs to the colon: An *in vitro* study using amylose coated tablets. *International Journal of Pharmaceutics.* 300(1-2): 89-94.

25. Husebye, E. (1995) Gastrointestinal motility disorders and bacterial overgrowth. *Journal of Internal Medicine.* 237: 419-427.

26. Mortazavi, S.M., M.R. Mohammadabadi, & M.R. Mozafari (2005) Applications and *in vivo* behaviour of lipid vesicles, in *Nanoliposomes: From Fundamentals to Recent Developments*, M.R. Mozafari & S.M. Mortazavi, Editors, Trafford Pub. Ltd, UK: 67-76.

27. Mozafari, M.R., E.T. Baran, S. Yurdugul, & A. Omri (2005) Liposome-based carrier systems, in *Nanoliposomes: From Fundamentals to Recent Developments*, M.R. Mozafari & S.M. Mortazavi, Editors, Trafford Pub. Ltd, UK: 67-76.

28. Awati, A., S.R. Konstantinov, B.A. Williams, A.D.L. Akkermans, M.W. Bosch, H. Smidt, & M.W.A. Verstegen (2005) Effect of substrate adaptation on the microbial

fermentation and microbial composition of faecal microbiota of weaning piglets studied *in vitro. Journal of the Science of Food and Agriculture*, 85(10): 1765-1772.

29. Coles, L.T., P.J. Moughan, & A.J. Darragh (2005) In vitro digestion and fermentation methods, including gas production techniques, as applied to nutritive evaluation of foods in the hindgut of humans and other simple-stomached animals. *Animal Feed Science and Technology*. 123-124(Part 1): 421-444.

30. Williams, B.A., M.W. Bosch, A. Awati, S.R. Konstantinov, H. Smidt, A.D.L. Akkermans, M.W.A. Verstegen, & S. Tamminga (2005) *In vitro* assessment of gastrointestinal tract (GIT) fermentation in pigs: Fermentable substrates and microbial activity. *Animal Research*. 54(3): 191-201.

31. Williams, B.A., M.W. Bosch, H. Boer, M.W.A. Verstegen, & S. Tamminga (2005) An *in vitro* batch culture method to assess potential fermentability of feed ingredients for monogastric diets. *Animal Feed Science and Technology*. 123-124(Part 1): 445-462.

Chapter 7

RECENT ADVANCES IN THE DELIVERY OF FOOD-DERIVED BIOACTIVES AND DRUGS USING MICROEMULSIONS

John Flanagan and Harjinder Singh
Riddet Centre, Massey University, Private Bag 11 222, Palmerston North, New Zealand.

Abstract: Microemulsions (so-called due to their small particle size; 5–100 nm) are thermodynamically stable, transparent, low viscosity and isotropic dispersions consisting of oil and water stabilized by an interfacial film of surfactant molecules. This chapter gives a short overview of the properties of microemulsions and gelled microemulsions, called organogels. Specific emphasis is placed on the ability of microemulsions and organogels to improve the solubilisation and bioavailability of food-derived bioactives and drugs, and recent advances in the food and pharmacological fields are presented.

Keywords: micelles; organogels; solubilisation; bioavailability

1. INTRODUCTION

The use of microemulsions as a delivery system has attracted a considerable degree of interest in the pharmaceutical sciences. This is due to the increased use of proteins and peptides for therapeutic purposes, and the ease with which proteins and peptides can be solubilised in microemulsions. In the food sciences, an increasing number of reports have detailed the bioactive capability of compounds isolated from varying food sources, and many of these compounds show significant health benefits when consumed in appropriate concentrations. However, some bioactive compounds exhibit poor solubility and low bioavailability, and delivery systems are being developed to help overcome these problems. Microemulsions, as a delivery system, may be a potential candidate to improve the solubility and increase the bioavailability of food-derived bioactive compounds [1].

M.R. Mozafari (ed.), Nanocarrier Technologies: Frontiers of Nanotherapy, 95–111.
© 2006 *Springer. Printed in the Netherlands.*

Microemulsions are thermodynamically stable, transparent isotropic solutions with particle sizes ranging from 5 to 100 nm, and arise from the spontaneous self-assembly of the hydrophobic or hydrophilic parts of surfactant molecules. Numerous studies have been conducted on micro-emulsions, researching their use in a wide variety of systems, including pharmaceutical, cosmetics, food, oil recovery, as models for biological membranes, and as reaction media, and new applications are constantly being reported.

Microemulsions are typically formed at exact concentrations of water, oil, surfactant, and possibly cosurfactant, and are deemed oil-in-water (o/w) or water-in-oil (w/o) depending on which is in the continuous phase. The concentrations at which microemulsions form are normally mapped out on ternary phase diagrams, similar to that shown in Fig. 7-1.

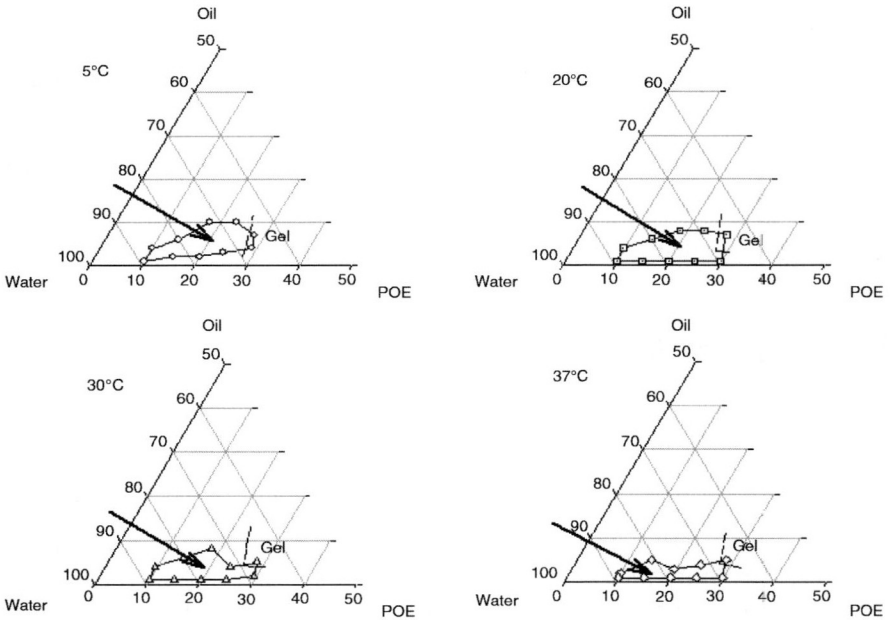

Figure 7-1. Typical ternary phase diagram for a soybean oil, polyoxyethylene ether surfactant (Brij 97) and water system at 5, 20, 30 and 37°C, showing areas of microemulsion formation. Reprinted with permission from Flanagan *et al.* 2006 [70].

Primarily, microemulsions differ from normal, coarse emulsions, in that microemulsions normally form spontaneously (no energy addition required), have very small particle sizes (<100 nm), are transparent/translucent and are thermodynamically stable. The oil type normally used in microemulsion formation is a hydrocarbon, or short- and medium-chain triglyceride. Long-chain triglycerides are more difficult to solubilise as they are semi-polar compared to hydrocarbon oils and they are too bulky to penetrate the interfacial film to assist in the formation of an optimal curvature [2].

Surfactants used for microemulsion formation may be divided into non-ionic, ionic, cationic and zwitterionic. Normally, non-ionic surfactants are preferred due to the reduced toxicity of these surfactants, compared to ionic or cationic surfactants [3]. Zwitterionic surfactants, such as the phospholipids, may also be used to formulate microemulsions. Surfactants which have been utilised in recent microemulsion studies are listed in Table 7-1. The ability of the surfactant to form a micelle is determined by its critical packing parameter (see Fig. 7-2) and also by its hydrophile-lipophile balance; these have previously been discussed in more detail [4]. In many cases, a cosurfactant, such as a short-chain alcohol, may be required to assist in microemulsion formation. The cosurfactant acts by further reducing the interfacial tension and also by partitioning between the surfactant molecules at the interface and changing the curvature of the interfacial layer. Other methods to "induce" microemulsion formation include the use of glycerol or polyglycerol in the water phase to reduce the polarity of the aqueous phase. Addition of sugars and salts may also aid in the formation of microemulsions due to their ability to hydrate the polar headgroups of the surfactant, thereby changing the curvature of the interfacial layer.

Under certain conditions, w/o microemulsions may form a gel, called an organogel. Gelatin is often used to induce gelation. Organogels may be exploited as delivery devices for hydrophobic and hydrophilic drugs and vaccines due to their ability to preserve the chemical and physical properties of the encapsulated material [5, 6, 7]. An important advantage of organogels over hydrogels is their ability to preserve the physical and chemical integrity of compounds solubilised within the microemulsion. In addition, microbial contamination of organogels is much less likely than contamination of hydrogels due to the presence of an organic continuous medium; furthermore, the aqueous domains are in the nanometre range, and hence orders of magnitude smaller than typical bacteria. Sorbitan monostearate organogels have also been studied as delivery devices for hydrophobic and hydrophilic drugs and vaccines [7-11]. Further information on lecithin [12] and gelatin organogels [13] has recently been provided.

Table 7-1. Examples of surfactants previously utilised in the formulation of microemulsions.

Surfactant	US 21CFR[a]	EU no.	Ref.
Lecithin and lecithin derivatives			
Pure phospholipid and mixed phospholipids	184.1400[b]	E322	[39, 70]
Hydroxylated phospholipids/lecithin	172.814[b]	E322	[41]
Glycerol fatty acid esters			
Polyglycerol fatty acid esters	172.854	E475	[65]
Propylene glycol fatty acid esters	172.856	E477	
Partial glycerides and derivatives			
Mono- and di-glycerides	184.1505[b]	E471	[43-45] [66]
Diacetyl tartaric acid esters of mono- and di-glycerides (DATEMS)	184.1101[b]		
Ethoxylated mono- and di-glycerides	172.834		[70]
Sucrose esters			
Mono-, di-, and tri-esters of sucrose with fatty acids	172.859	E473	[26, 67]
Sorbitan fatty acid esters			
Sorbitan monostearate	172.842	E491	[9-11]
Sorbitan monolaurate		E493	
Sorbitan monooleate		E494	
Polyoxyethylene sorbitan fatty acid esters			
POE sorbitan monostearate[*]	172.836	E435	[68]
POE sorbitan tristearate[*]	172.838	E436	
POE sorbitan monooleate[*]	172.840	E433	
Other			
Propylene glycol ethers	184.1666[b]		[69]
Sodium lauryl sulphate/dodecyl sulphate	172.822		[30]
Sodium bis(ethylhexyl) sulfosuccinate			[37]
Cetyltrimethyl ammonium bromide (CTAB)			
Polyoxyethylene ethers (POE)			
Fructose esters			[70]

[a] United States Code of Federal Regulations, Title 21, Volume 3, Revised as of April 1, 2004, from the U.S. Government Printing Office. Accessed via http://www.gpoaccess.gov/

[b] Generally recognised as safe (GRAS)

[*] Mixtures of POE sorbitan fatty acid esters are generally used to formulate microemulsions

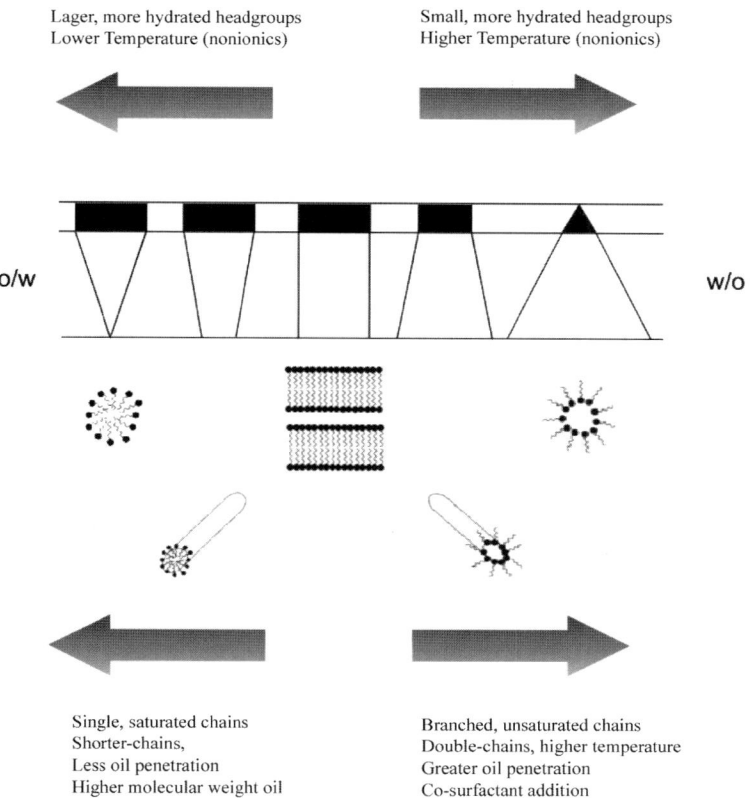

Larger, more hydrated headgroups
Lower Temperature (nonionics)

Small, more hydrated headgroups
Higher Temperature (nonionics)

o/w

w/o

Single, saturated chains
Shorter-chains,
Less oil penetration
Higher molecular weight oil

Branched, unsaturated chains
Double-chains, higher temperature
Greater oil penetration
Co-surfactant addition

Figure 7-2. Effect of solution conditions on the headgroup size and resulting effect on the surfactant molecule shape and arrangement. Reprinted from Advanced Drug Delivery Reviews, Vol 45, Lawrence and Rees, Microemulsion-based media as novel drug delivery systems, Pages 89-121, Copyright (2000), with permission from Elsevier. Originally reprinted from Colloids and Surfaces A, Vol. 91, Israelachvili, The science and application of emulsions – an overview, Pages 1-8, Copyright (1994), with permission from Elsevier.

In a recent review Bagwe *et al.,* [5] concluded that lecithin organogels offer superior advantages over gelatin organogels due to the biocompatibility of lecithin formulations and the relative ease with which they have been used. Biocompatible lecithin organogels have now been reported [14].

Recent reviews on microemulsions have focussed on their use in drug delivery [3, 5] and applications in foods [1, 2]. This review will focus on recent advances in the solubilisation of bioactives and drugs in micro-emulsions, including studies which have reported on the resulting bioavail-ability of the solubilised material.

2.   SOLUBILISATION AND BIOAVAILABILITY OF FOOD-DERIVED BIOACTIVES USING MICROEMULSIONS

The importance of certain food constituents is being realised and much research has centred on the identification, separation and in-vitro determination of ability of bioactives to modulate specific health effects. Realisation that many of these bioactives are susceptible to breakdown (thermal or light induced oxidation, proteolysis in the stomach, or interaction with other food constituents), has prompted investigation of delivery systems to protect the bioactives from these detrimental effects. One such delivery system is microemulsions. Many different bioactive compounds have been solubilised in microemulsions, and increased anti-oxidative effects, bioavailability and solubility, among other effects, have been reported. These are listed in Table 7-2, and discussed in more detail below.

2.1   Recent advances in applications of microemulsions in food

A considerable body of work has focussed on the ability of microemulsions to protect against oxidation. In a series of studies, the antioxidative effect of flavanols and their glycosides [15], green tea polyphenols [16], and resveratrol [17] in sodium dodecyl sulphate (SDS) and cetyl trimethylammonium bromide (CTAB) micelles, alone or in combination with α-tocopherol were investigated. All systems studied proved to have antioxidative effects, but this depended on the hydrophilicity of the compounds and the microenvironment of the reaction media. In addition, it was shown that green tea polyphenols could regenerate α-tocopherol from its oxidised state while in SDS micelles [18].

The solubilisation and protection of Vitamin E against oxidation in microemulsions prepared with POE surfactants was reported [19]. Patents have also been filed for the protection of flavour or aroma [20, 21] and vitamins [22] using microemulsion technology. A separate study found that β-carotene, solubilised in a SDS w/o microemulsion, was more stable against oxidation than in an o/w or bicontinuous microemulsion system [23]. The authors postulated that the hydrophobic barrier of the SDS association structures in w/o microemulsions can hinder electron transfer from β-carotene and hence make their oxidation difficult.

Table 7-2. Recent studies involving solubilisation of food-derived components in microemulsions. Abbreviations: 3G1O – triglycerol monooleate; AOT – sodium bis(ethylhexyl) sulfosuccinate; CTAB – cetyl trimethylammonium bromide; EM – ethoxylated mono-/di-glycerides; POE – polyoxyethylene ether; POE-SE – polyoxyethylene sorbitan fatty acid ester; SDS – sodium dodecyl sulphate; SMS – sucrose monostearate.

Component Solubilised	Surfactant(s) employed	Results of study	Ref.
α-tocopherol and flavanols	SDS, CTAB	Antioxidative	[15]
α-tocopherol and resveratrol	SDS, CTAB	Antioxidative	[17]
α-tocopherol and green tea polyphenols	SDS, CTAB	Antioxidative	[16, 18]
β-Carotene	SDS	Stability against oxidation	[23]
β-Carotene	POE-SE, SMS	Increased bioavailability	[24]
Lycopene	POE-SS, EM, 3G1O, SMS	Increased solubility	[25]
Lycopene	POE-SE	Increased light stability	[26]
Lutein	POE-SE	Increased solubility	[27]
Phytosterol	POE-SE	Increased solubility	[28]
Carrageenan	POE	Increased skin permeation	[48]
Immunoglobulin G	AOT	Solubilisation, change of microemulsion structure	[29]
α-Lactalbumin	AOT	Solubilisation	[30]

Increased bioavailability of β-carotene was observed when administered in a microemulsion form (using a mixture of sorbitan monostearate surfactant and its polyoxyethylene form), compared to a conventional dispersion [24]. The solubilisation capacity of lycopene (an essential carotenoid sensitive to light degradation) in w/o, bicontinuous and o/w microemulsions was increased 4 times, compared to solubility in an essential oil, limonene. Further increases in lycopene solubility were observed when mixtures of surfactants were used [25]. In a subsequent study, the light stability of lycopene was found to be improved following solubilisation of lycopene in Tween-based microemulsions compared to solubilisation in acetone [26]. Garti *et al.* [26] have also reported the encapsulation of a bromine-based bactericide in the presence of butyl lactate into microemulsions prepared with an ethoxylated surfactant, Brij 96. The authors claimed that the microemulsion systems used to encapsulate bactericides could also be used to encapsulate pesticides [26].

Up to 4% (w/w) lutein (a naturally occurring, poorly soluble, carotenoid associated with reduced risks of cataract development) was incorporated into the bicontinuous phase of microemulsions containing R-(+)-limonene, ethanol, glycerol, Tween 80 and water [27]. The solubilisation capacity of phytosterol was increased 6-fold following incorporation in a microemulsion formulation, compared to the solubility of phytosterol in limonene [28].

The inclusion of proteins in microemulsions has also been the basis for a number of research topics. Proteins hosted in microemulsions may find application in research into enzyme activity and protein separation; they can alter surfactant self-association and phase behaviour and can promote the formation of novel solvents and materials [29].

Recently, the incorporation of immunoglobulin G [29] and α-lactalbumin [30] into microemulsion systems has been reported. However, a further study has shown that the conformation of immunoglobulin G entrapped in a w/o microemulsion changed over time as it partitioned towards the interface [31]. This highlights the fact that protein denaturation may be a very important issue during the solubilisation of protein and peptide drugs within microemulsions. A significant increase in the volume of research conducted in the inclusion of proteins into microemulsion systems may be expected in the near future.

It appears that very little work has been conducted on the bio-availabilities of food-derived bioactives following their solubilisation in microemulsions. With the increased importance of bioactives in today's health-conscious consumer society, it is expected that research in this area will grow considerably.

3. SOLUBILISATION AND BIOAVAILABILITY OF DRUGS USING MICROEMULSIONS AND ORGANOGELS

The development of numerous recombinant protein drugs such as hormones and vaccines for therapeutic purposes has thrown up some problems with their delivery. These protein drugs often are high molecular weight proteins with poor solubility and may be sensitive to environmental conditions. Oral delivery of protein drugs is severely hampered by their exposure to acidic and proteolytic environment of the gastrointestinal tract. Other delivery routes, such as dermal, ocular, parenteral, may also be used to deliver proteins drugs, however systemic administration in high doses can lead to side effects and can also be toxic. Microemulsions may hold the solution to many oral delivery problems. They can be used to solubilise poorly soluble peptides/proteins and can possibly aid in absorption. Alternatively, microemulsions in a gelled form (organogels) may also be used in a dermal delivery system with increased permeability and controlled release properties. The potential of using microemulsions to enhance absorption of peptides and protein drugs through dermal, peroral, parenteral and ocular routes are frequently discussed in drug delivery reviews [32, 33, 34].

Details of recent findings relating to the solubilisation and delivery (oral, dermal, ocular, and parenteral) of drugs (including protein and peptide drugs) are outlined below.

3.1 Recent advances in applications of microemulsions in pharmaceutics

Djordjevic *et al.* [35] reported the solubilisation of diclofenac diethylamine in microemulsions formulated with water, isopropyl myristate and either caprylic glycerides or polyglcerol-6 dioleate as surfactant. Both lipophilic (abamectin) and hydrophilic (levamisole phosphate) drugs were solubilised together in a microemulsion designed for injectable formulations [36]. The microemulsions were prepared using mono- and diglycerides as surfactant. Malcolmson *et al.,* reported the solubilisation of testosterone propionate [37] and testosterone enanthate [38] in microemulsion formulations.

The effects of adding two sparingly soluble drugs (Felodipine and BIBP3226) on the phase behaviour and microstructure of the resulting microemulsions were examined [39]. Subsequently, a pharmaceutically acceptable microemulsion system, comprising of medium-chain triglycerides, a combination of soybean phosphatidyl choline and hydroxystearate polyethylene glycol as surfactants, and polyethylene glycol and ethanol as cosurfactants was developed [40]. The authors claimed that the microemulsion could be administered by intravenous infusion up to a dose of 0.5 mL/kg without producing any significant effect on the acid-base balance, blood gases, plasma electrolytes or heart rate of rats. A further study found decreased toxicity levels in mice which were injected with a microemulsion system containing Amphotericin (a potent fungistatic and fungicidal antibiotic) compared to the commercially available mixed micellar dispersion of Amphotericin [41].

Initial studies suggested that microemulsions could be used to enhance the oral bioavailability of drugs, including peptides [42, 43]. Subsequent studies substantiated this claim. Intraduodenal bioavailability of a water-soluble peptide antagonist was increased from 1% in saline solution to 29% in a microemulsion formulation [44]. Further studies on Calcein, a poorly absorbed molecular marker, showed an increase in absorption from 2% in aqueous solution to 37% in a w/o microemulsion after intraduodenal administration in anaesthetised rats [45]. Lyons *et al.* [46] found that, following intraduodenal administration to rats, the bioavailability of a muramyl dipeptide increased up to 10-fold when solubilised in w/o micro-emulsions, compared to aqueous solution. Microemulsions have also been utilized for increasing dermal delivery of both hydrophobic and hydrophilic compounds [47, 48].

The enhancement of absorption activity observed in microemulsion-delivered compounds is dependent on the type of emulsifying agent, particle size of the dispersed phase (if absorption is dissolution rate limited), pH, solubility of the compound, and type of lipid phase used [32]. Different hypotheses have been put forward to explain the increased bioavailability of compounds solubilised within microemulsions. It was suggested [43, 45, 46] that the presence of medium-chain triglycerides was the main reason for increased bioavailability as medium-chain triglycerides are known as absorption enhancing agents. Monoglycerides, phospholipids and fatty acids, usually present in microemulsion formulations, are also known to cause damage to the gastrointestinal epithelium and result in increased permeability [49]. Alternatively, the presence of high concentrations of surfactant could also damage the gastrointestinal epithelium thus promoting permeability.

Research has been carried out on the use of microemulsions and organogels as delivery systems for the controlled release and increased bioavailability of drugs and has been reviewed recently by Kreilgaard [47]. Chen *et al.* [50] investigated the *in vitro* permeation of triptolide using Franz diffusion cells fitted with mouse skins. Triptolide has been found to pos-sess many valuable functions, such as immunosuppressive, anti-cancer and antifertility activities. Tiptolide-loaded microemulsions showed controlled, sustained and prolonged delivery with reduced skin irritancy, compared to an aqueous solution of 20% propylene glycol containing 0.025% triptolide. A separate study found that the in vivo transdermal penetration of lidocaine, used to induce analgesia, was improved when in a microemulsion formulation, compared to a commercial cream [51]. The same authors also attempted to develop a microemulsion system for transdermal delivery of paracetamol for paediatric use [52]. *In vivo* and *in vitro* studies showed positive results; however, the authors recognised the need for longer pharmacokinetic studies in conscious animals.

Gelatin-containing microemulsion-based organogels were formulated to encapsulate salicylate (which possesses anti-inflammatory, analgesic, and antipyretic properties), and iontophoresis was used to facilitate their transfer across a dermal barrier resulting in substantially higher release rates compared to passive diffusion [13].

Gelled microemulsions have also been reviewed as a delivery system for vaginal spermicides, exhibiting potent contraceptive activity and absence of local, systemic or reproductive toxicity [6].

Fialho & da Silva-Cunha [53] attempted to improve the ocular bioavailability of dexamethasone by solubilisation in an o/w microemulsion. They reported a microemulsion formulation with good stability and no evidence to suggest irritancy, along with increased penetration. In addition, the microemulsion formulation was found to give a longer release compared to a conventional preparation. Rupenthal *et al.* [54] have also reported on the ocular delivery of antisense oligonucleotides using w/o microemulsion formulations.

Research has also been conducted on the use of microemulsions for parenteral administration of drugs. The research group of Kim at the Seoul National University have developed microemulsion systems for the parenteral delivery of flurbiprofen [55, 56] and all-trans-retinoic acid [57].

The research group at the University of Otago, New Zealand, have developed a method for preparing nanocapsules by interfacial polymerisation of w/o microemulsions [58, 59]. They have used these nanocapsules to develop systems for delivery of insulin, without restricting the *in vivo* or *in vitro* bioactivity of insulin [60].

A recent study has investigated the possibility of using artificial neural network data modelling to design microemulsions for oral delivery of a combination of antitubercular drugs [61]. A tocopheryl polyethylene glycol 1000 succinate surfactant has also been studied as a means for formulating microemulsions which may be suitable for oral delivery of protein drugs [62]. This is a major step forward in terms of use of microemulsions in the pharmaceutical field, and further research in this area can be expected.

Microemulsions have also found application in the cosmetics industry. The incorporation of Jojoba oil (which may have skin health properties) into microemulsions has been reported [63] and a patent has recently been filed by Unilever outlining the use of microemulsions as an antiperspirant, formulated as either a gel or as a spray, containing inorganic salts, cosmetically acceptable oils, and a quartenary and non-ionic surfactant [64].

4.   SUMMARY AND FUTURE OUTLOOK

Microemulsions have been shown to increase the solubilities of poorly soluble food-derived bioactives and drugs. In addition, the bioavailabilities of these drugs have been shown to be increased following delivery by a number of routes (dermal, oral, ocular). However, claims of increased bioavailability of food-derived bioactives need to be substantiated with in vivo animal and human trials, and much more work may be expected in this area in the near future.

5.   REFERENCES

1. Flanagan, J. & Singh, H. (2006). Microemulsions: A potential delivery system for bioactives in food. *Critical Reviews in Food Science and Nutrition,* 46(3), 221-237.
2. Gaonkar, A. G. & Bagwe, R. P. (2003). Microemulsions in Foods: Challenges and Applications. *Surfactant Science Series,* (109), 407-430.
3. Tenjarla, S. (1999). Microemulsions: An overview and pharmaceutical applications. *Critical Reviews in Therapeutic Drug Carrier Systems,* 16(5), 461-521.
4. Lawrence, M. J. & Rees, G. D. (2000). Microemulsion-based media as novel drug delivery systems. *Advanced Drug Delivery Reviews,* 45(1), 89-121.
5. Bagwe, R. P., Kanicky, J. R., Palla, B. J., Patanjali, P. K., & Shah, D. O. (2001). Improved drug delivery using microemulsions: Rationale, recent progress and new horizons. *Critical Reviews in Therapeutic Drug Carrier Systems,* 18(1), 77-140.
6. D'Cruz, O. J. & Uckun, F. M. (2001). Gel-microemulsions as vaginal spermicides and intravaginal drug delivery vehicles. *Contraception,* 64(2), 113-123.
7. Murdan, S., Gregoriadis, G., & Florence, A. T. (1999). Novel sorbitan monostearate organogels. *Journal of Pharmaceutical Sciences,* 88(6), 608-614.
8. Murdan, S., Gregoriadis, G., & Florence, A. T. (1996). Non-ionic surfactant based organogels incorporating niosomes. *Stp Pharma Sciences,* 6(1), 44-48.
9. Murdan, S., Gregoriadis, G., & Florence, A. T. (1999). Inverse toroidal vesicles: precursors of tubules in sorbitan monostearate organogels. *International Journal of Pharmaceutics,* 183(1), 47-49.

10. Murdan, S., Gregoriadis, G., & Florence, A. T. (1999). Sorbitan monostearate polysorbate 20 organogels containing niosomes: a delivery vehicle for antigens? *European Journal of Pharmaceutical Sciences,* 8(3), 177-185.

11. Murdan, S., van den Bergh, B., Gregoriadis, G., & Florence, A. T. (1999). Water-in-sorbitan monostearate organogels (water-in-oil gels). *Journal of Pharmaceutical Sciences,* 88(6), 615-619.

12. Shchipunov, Y. A. (2001). Lecithin organogel - A micellar system with unique properties. *Colloids and Surfaces A Physicochemical and Engineering Aspects,* 183, 541-554.

13. Kantaria, S., Rees, G. D., & Lawrence, M. J. (1999). Gelatin-stabilised microemulsion-based organogels: rheology and application in iontophoretic transdermal drug delivery. *Journal of Controlled Release,* 60(2-3), 355-365.

14. Angelico, R., Ceglie, A., Colafemmina, G., Lopez, F., Murgia, S., Olsson, U., & Palazzo, G. (2005). Biocompatible lecithin organogels: Structure and phase equilibria. *Langmuir,* 21(1), 140-148.

15. Zhou, B., Miao, Q., Yang, L., & Liu, Z. L. (2005). Antioxidative effects of flavonols and their glycosides against the free-radical-induced peroxidation of linoleic acid in solution and in micelles. *Chemistry-A European Journal,* 11(2), 680-691.

16. Zhou, B., Jia, Z. S., Chen, Z. H., Yang, L., Wu, L. M., & Liu, Z. L. (2000). Synergistic antioxidant effect of green tea polyphenols with alpha-tocopherol on free radical initiated peroxidation of linoleic acid in micelles. *Journal of the Chemical Society-Perkin Transactions 2,* 4, 785-791.

17. Fang, J. G., Lu, M., Chen, Z. H., Zhu, H. H., Li, Y., Yang, L., Wu, L. M., & Liu, Z. L. (2002). Antioxidant effects of resveratrol and its analogues against the free-radical-induced peroxidation of linoleic acid in micelles. *Chemistry-A European Journal,* 8(18), 4191-4198.

18. Zhou, B., Wu, L. M., Yang, L., & Liu, Z. L. (2005). Evidence for alpha-tocopherol regeneration reaction of green tea polyphenols in SDS micelles. *Free Radical Biology and Medicine,* 38(1), 78-84.

19. Chiu, Y. C. & Yang, W. L. (1992). Preparation of vitamin E microemulsion possessing high resistance to oxidation. *Colloids and Surfaces,* 63, 311-322.

20. Chmiel, O., Traitler, H., & Voepel, K. (1997). Food microemulsion formulations. US Patent No. 5,674,549.

21. Chung, S. L., Tan, C.-T., Tuhill, I. M., & Scharpf, L. G. (1994). Transparent oil-in-water microemulsion flavor or fragrance concentrate, process for preparing same, mouthwash or perfume composition containing said transparent microemulsion concentrate, and process for preparing same. US Patent No. 5,283,056.

22. Bauer, K., Neuber, C., Schmid, A., & Volker, K. M. (2002). Oil in water microemulsion. US Patent No. 6,426,078.

23. Szymula, M. (2004). Atmospheric oxidation of beta-carotene in aqueous, pentanol, SDS microemulsion systems in the presence and absence of vitamin C. *Journal of Dispersion Science and Technology,* 25(2), 129-137.

24. Van den Braak, M., Szymula, M., & Ford, M. A. (2001). Stable, optically clear compositions. US Patent No. 6,251,441.

25. Spernath, A., Yaghmur, A., Aserin, A., Hoffman, R. E., & Garti, N. (2002). Food-grade microemulsions based on nonionic emulsifiers: Media to enhance lycopene solubilization. *Journal of Agricultural and Food Chemistry,* 50(23), 6917-6922.

26. Garti, N., Yaghmur, A., Aserin, A., Spernath, A., Elfakess, R., & Ezrahi, S. (2004). Solubilization of active molecules in microemulsions for improved environmental protection. *Colloids and Surfaces A-Physicochemical and Engineering Aspects,* 230(1-3), 183-190.

27. Amar, I., Aserin, A., & Garti, N. (2003). Solubilization patterns of lutein and lutein esters in food grade nonionic microemulsions. *Journal of Agricultural and Food Chemistry,* 51(16), 4775-4781.

28. Spernath, A., Yaghmur, A., Aserin, A., Hoffman, R. E., & Garti, N. (2003). Self-diffusion nuclear magnetic resonance, microstructure transitions, and solubilization capacity of phytosterols and cholesterol in Winsor IV food-grade microemulsions. *Journal of Agricultural and Food Chemistry,* 51(8), 2359-2364.

29. Gerhardt, N. I. & Dungan, S. R. (2002). Time-dependent solubilization of IgG in AOT-brine-isooctane microemulsions: Role of cluster formation. *Biotechnology and Bioengineering,* 78(1), 60-72.

30. Rohloff, C. M., Shimek, J. W., & Dungan, S. R. (2003). Effect of added alpha-lactalbumin protein on the phase behavior of AOT-brine-isooctane systems. *Journal of Colloid and Interface Science,* 261(2), 514-523.

31. Gerhardt, N. I. & Dungan, S. R. (2004). Changes in microemulsion and protein structure in IgG-AOT-brine isooctane systems. *Journal of Physical Chemistry B,* 108(28), 9801-9810.

32. Sood, A. & Panchagnula, R. (2001). Peroral route: An opportunity for protein and peptide drug delivery. *Chemical Reviews,* 101(11), 3275-3303.

33. Vandamme, T. F. (2002). Microemulsions as ocular drug delivery systems: recent developments and future challenges. *Progress in Retinal and Eye Research,* 21(1), 15-34.

34. de Oliveira, A. G., Scarpa, M. V., Correa, M. A., Cera, L. F. R., & Formariz, T. P. (2004). Microemulsions: Structure and application as drug delivery systems. *Quimica Nova,* 27(1), 131-138.

35. Djordjevic, L., Primorac, M., Stupar, M., & Krajisnik, D. (2004). Characterization of caprylocaproyl macrogolglycerides based microemulsion drug delivery vehicles for an amphiphilic drug. *International Journal of Pharmaceutics,* 271(1-2), 11-19.

36. Sari, P., Razzak, M., & Tucker, I. G. (2004). Isotropic systems of medium-chain mono- and diglycerides for solubilization of lipophilic and hydrophilic drugs. *Pharmaceutical Development and Technology,* 9(1), 97-106.

37. Malcolmson, C., Satra, C., Kantaria, S., Sidhu, A., & Lawrence, M. J. (1998). Effect of oil on the level of solubilization of testosterone propionate into nonionic oil-in-water microemulsions. *Journal of Pharmaceutical Sciences,* 87(1), 109-116.

38. Malcolmson, C., Barlow, D. J., & Lawrence, M. J. (2002). Light-scattering studies of testosterone enanthate containing soybean oil/C18:1E10/water oil-in-water micro-emulsions. *Journal of Pharmaceutical Sciences,* 91(11), 2317-2331.

39. von Corswant, C. & Thoren, P. E. G. (1999). Solubilization of sparingly soluble active compounds in lecithin-based microemulsions: Influence on phase behavior and microstructure. *Langmuir,* 15(11), 3710-3717.

40. von Corswant, C., Thoren, P., & Engstrom, S. (1998). Triglyceride-based microemulsion for intravenous administration of sparingly soluble substances. *Journal of Pharmaceutical Sciences,* 87(2), 200-208.

41. Brime, B., Moreno, M. A., Frutos, G., Ballesteros, M. P., & Frutos, P. (2002). Amphotericin B in oil-water lecithin-based microemulsions: Formulation and toxicity evaluation. *Journal of Pharmaceutical Sciences,* 91(4), 1178-1185.

42. Constantinides, P. P., Scalart, J. P., Lancaster, C., Marcello, J., Marks, G., Ellens, H., & Smith, P. L. (1994). Formulation and intestinal-absorption enhancement evaluation of water-in-oil microemulsions incorporating medium-chain glycerides. *Pharmaceutical Research,* 11(10), 1385-1390.

43. Constantinides, P. P., Lancaster, C. M., Marcello, J., Chiossone, D. C., Orner, D., Hidalgo, I., Smith, P. L., Sarkahian, A. B., Yiv, S. H., & Owen, A. J. (1995). Enhanced intestinal-absorption of an RGD peptide from water-in-oil microemulsions of different composition and particle-size. *Journal of Controlled Release,* 34(2), 109-116.

44. Constantinides, P. P. (1995). Lipid microemulsions for improving drug dissolution and oral absorption: Physical and biopharmaceutical aspects. *Pharmaceutical Research,* 12(11), 1561-1572.

45. Constantinides, P. P., Welzel, G., Ellens, H., Smith, P. L., Sturgis, S., Yiv, S. H., & Owen, A. B. (1996). Water-in-oil microemulsions containing medium-chain fatty acids salts: Formulation and intestinal absorption enhancement evaluation. *Pharmaceutical Research,* 13(2), 210-215.

46. Lyons, K. C., Charman, W. N., Miller, R., & Porter, C. J. H. (2000). Factors limiting the oral bioavailability of N-acetylglucosaminyl-N-acetylmuramyl dipeptide (GMDP) and enhancement of absorption in rats by delivery in a water-in-oil microemulsion. *International Journal of Pharmaceutics,* 199(1), 17-28.

47. Kreilgaard, M. (2002). Influence of microemulsions on cutaneous drug delivery. *Advanced Drug Delivery Reviews,* 54, S77-S98.

48. Valenta, C. & Schultz, K. (2004). Influence of carrageenan on the rheology and skin permeation of microemulsion formulations. *Journal of Controlled Release,* 95(2), 257-265.

49. Ilback, N. G., Nyblom, M., Carlfors, J., Fagerlund-Aspenstrom, B., Tavelin, S., & Glynn, A. W. (2004). Do surface-active lipids in food increase the intestinal permeability to toxic substances and allergenic agents? *Medical Hypotheses,* 63(4), 724-730.

50. Chen, H. B., Chang, X. L., Weng, T., Zhao, X. Z., Gao, Z. H., Yang, Y. J., Xu, H. B., & Yang, X. L. (2004). A study of microemulsion systems for transdermal delivery of triptolide. *Journal of Controlled Release,* 98(3), 427-436.

51. Sintov, A. C. & Shapiro, L. (2004). New microemulsion vehicle facilitates percutaneous penetration in vitro and cutaneous drug bioavailability *in vivo*. *Journal of Controlled Release*, 95(2), 173-183.

52. Sintov, A. C., Krymberk, I., Gavrilov, V., & Gorodischer, R. (2003). Transdermal delivery of paracetamol for paediatric use: effects of vehicle formulations on the percutaneous penetration. *Journal of Pharmacy and Pharmacology*, 55(7), 911-919.

53. Fialho, S. L. & Silva-Cunha, A. (2004). New vehicle based on a microemulsion for topical ocular administration of dexamethasone. *Clinical and Experimental Ophthalmology*, 32(6), 626-632.

54. Rupenthal, I. D., Green, C. R., & Alany, R. G. (2005). Stability and ocular delivery of antisense oligonucleotides using water-in-oil microemulsions. Proceedings of Formulation and Delivery of Bioactives Conference, Dunedin. 18-19th February, 2005.

55. Park, K. M., Lee, M. K., Hwang, K. J., & Kim, C. K. (1999). Phospholipid-based microemulsions of flurbiprofen by the spontaneous emulsification process. *International Journal of Pharmaceutics*, 183(2), 145-154.

56. Park, K. M. & Kim, C. K. (1999). Preparation and evaluation of flurbiprofen-loaded microemulsion for parenteral delivery. *International Journal of Pharmaceutics*, 181(2), 173-179.

57. Hwang, S. R., Lim, S. J., Park, J. S., & Kim, C. K. (2004). Phospholipid-based microemulsion formulation of all-trans-retinoic acid for parenteral administration. *International Journal of Pharmaceutics*, 276(1-2), 175-183.

58. Pitaksuteepong, T., Davies, N. M., Tucker, I. G., & Rades, T. (2002). Factors influencing the entrapment of hydrophilic compounds in nanocapsules prepared by interfacial polymerisation of water-in-oil microemulsions. *European Journal of Pharmaceutics and Biopharmaceutics*, 53(3), 335-342.

59. Watnasirichaikul, S., Rades, T., Tucker, I. G., & Davies, N. M. (2002). Effects of formulation variables on characteristics of poly (ethylcyanoacrylate) nanocapsules prepared from w/o microemulsions. *International Journal of Pharmaceutics*, 235(1-2), 237-246.

60. Watnasirichaikul, S., Rades, T., Tucker, I. G., & Davies, N. M. (2002). In-vitro release and oral bioactivity of insulin in diabetic rats using nanocapsules dispersed in biocompatible microemulsion. *Journal of Pharmacy and Pharmacology*, 54(4), 473-480.

61. Agatonovic-Kustrin, S., Glass, B. D., Wisch, M. H., & Alany, R. G. (2003). Prediction of a stable microemulsion formulation for the oral delivery of a combination of antitubercular drugs using ANN methodology. *Pharmaceutical Research*, 20(11), 1760-1765.

62. Ke, W. T., Lin, S. Y., Ho, H. O., & Sheu, M. T. (2005). Physical characterizations of microemulsion systems using tocopheryl polyethylene glycol 1000 succinate (TPGS) as a surfactant for the oral delivery of protein drugs. *Journal of Controlled Release*, 102(2), 489-507.

63. Shevachman, M., Shani, A., & Garti, N. (2004). Formation and investigation of microemulsions based on Jojoba oil and nonionic surfactants. *Journal of the American Oil Chemists Society,* 81(12), 1143-1152.

64. Ma, Z. & Brucks, R. M. (2004). Antiperspirant compositions comprising micro-emulsions., US Patent No. 6,790,435.

65. Kawakami, K., Yoshikawa, T., Moroto, Y., Kanaoka, E., Takahashi, K., Nishihara, Y., & Masuda, K. (2002). Microemulsion formulation for enhanced absorption of poorly soluble drugs - I. Prescription design. Journal of Controlled Release, 81(1-2), 65-74.

66. Bagwe, R. P. & Shah, D. O. (2002). Effect of various additives on solubilization, droplet size and viscosity of canola oil in oil-in-water food grade microemulsions. *Abstracts of Papers, 223rd ACS National Meeting, April 7-11, Orlando, Florida.*

67. Glatter, O., Orthaber, D., Stradner, A., Scherf, G., Fanun, M., Garti, N., Clement, V., & Leser, M. E. (2001). Sugar-ester nonionic microemulsion: Structural characterization. *Journal of Colloid and Interface Science*, 241(1), 215-225.

68. Yaghmur, A., Aserin, A., Antalek, B., & Garti, N. (2003). Microstructure considerations of new five-component Winsor IV food-grade microemulsions studied by pulsed gradient spin-echo NMR, conductivity, and viscosity. *Langmuir*, 19(4), 1063-1068.

69. Ghoulam, M. B., Moatadid, N., Graciaa, A., & Lachaise, J. (2004). Quantitative effect of nonionic surfactant partitioning on the hydrophile-lipophile balance temperature. *Langmuir*, 20, 2584-2589.

70. Flanagan, J., Kortegaard, K., Pinder, D. N., Rades, T., & Singh, H. (2006). Solubilisation of soybean oil in microemulsions using various surfactants. *Food Hydrocolloids*, 20(2-3), 253-260.

Chapter 8

PHARMACOKINETIC MODULATION WITH PARTICULATE DRUG FORMULATIONS

Marek Langner [1] and Arkadiusz Kozubek [2]

[1]*Institute of Physics, Wroclaw Technical University, Wyb. Wyspianskiego 27, 50-370 Wroclaw, Poland;* [2]*Institute of Biochemistry and Molecular Biology, University of Wroclaw, Przybyszewskiego 63/77, 51-148 Wroclaw, Poland*

Abstract: Pharmacokinetics and biodistribution, along with biological activity, are parameters which determine the quality of pharmacological therapy. The search for improved active compound specificity has been continued for decades. Initially it was believed that the chemical modification of active compounds will suffice to improve their performance. In the course of the ever increasing complexity of the designed drug molecules, however, they still have not acquired all the desired parameters. In parallel, the concept of supramolecular aggregates as active compound carriers has been developed and its applicability explored. Initial failures, which resulted from too simplistic definitions of the problem, have been followed by the discovery of long-circulating liposomes and increasing knowledge of supramolecular aggregate interaction with biological structures and body defense systems. All these have resulted in a growing number of concepts and ideas that caused a spur of applications of particulate drug formulations. All these developments were intended to improve the performance of the active compound in vivo by altering its concentration profile in time, the optimalization of its biodistribution, and its specificity towards selected tissues or even cells. The progress in these areas has enforced the rethinking of our perception of the drug itself, its interactions with biological structures, and the body as a whole. In this chapter, selected issues related to targeted drug delivery systems are presented, discussed, and assesed within the context of recent developments in the pharmacology of supramolecular aggregates and related fields.

Key words: Supramolecular aggregates, biodistribution, targeted drug delivery systems, pharmacokinetics.

M.R. Mozafari (ed.), Nanocarrier Technologies: Frontiers of Nanotherapy, 113–138.

1.   INTRODUCTION

Historically, the drug discovery process has in principle been a trial and error approach in different variations. This was due to limited knowledge on the molecular mechanisms of the targeted diseases, the lack of ability to identify relevant pharmacological targets, and difficulties with designing the chemical structure of active compounds that would ensure optimal interaction with those targets. In the mid of the last century, the elements of rational drug design were implemented as an integral part of research and the development process. When knowledge of molecular processes had increased sufficiently, identifying molecular targets and implementing procedures for designing the chemical structure of drugs became possible and effective. At the same time, however, relatively little effort has been given to the performance of investigated compounds in reaching target cells/tissues. This resulted in the successful development and synthesis of a large number of potent drug candidates, of which a substantial fraction had to be later rejected due to their poor pharmacokinetic performance and/or undesired biodistribution [45]. In fact, for a long time only experimental methods were available for providing information on compound behavior *in vivo*. Difficulties with these costly, time consuming, and ethically questionable studies using animal models stimulated research towards the development of *in vitro* tests and methods for selecting compounds in accordance to their ability to reach targeted cells and subsequently intracellular regions. Physicochemical (PAMPA; potentiometric; water/octanol, and water/membrane partition coefficients) and cellular (Caco-2 monolayer) tests for evaluating a compound's ability to cross the plasma membrane have been developed [9, 62, 120, 121]. In parallel, methodologies capable of estimating a compound's susceptibility for metabolic hydrolysis and/or modifications have been devised [128, 191]. At the same time, intensive research has been carried out to characterize compounds when still in the design stage. Parameters useful for predicting how efficiently compounds passively cross the lipid bilayer barrier have been devised - the water/octanol partitioning coefficient, Lipinski's principle, topological indexes, polar surface area and similarity/dissimilarity criteria are now commonly used [53, 126, 180]. All these methodologies assume that chemical modification of a compound suffices to obtain the desired pharmacokinetic parameters without loosing its biological activity. So far this belief has found very little empirical justification. Consequently, despite devising an increasing number of parameters and criteria, their predictive power is still very limited and applicable mainly to a selected group of compounds. The major problem with this type of approach is the way a compound's interaction with "biological space" is perceived. Perhaps, for conceptual reasons, it is better to treat a compound as a carrier of a certain type of information, which is then utilized to perform certain tasks within the biological system [172]. The amount of such information encoded within a single molecular structure should reflect the complexity of tasks it is intended to perform. In other words, the amount of information reflects the, extent of simplification that has to be made in order to describe the biological system being treated. The most frequently tested and evaluated parameters, before the first experiments *in vivo* are carried out, are biological activity and the ability to passively cross the plasma membrane. This means that the biological system is in fact reduced to three elements: extracellular space, intracellular space, and the target protein (Fig. 8-1).

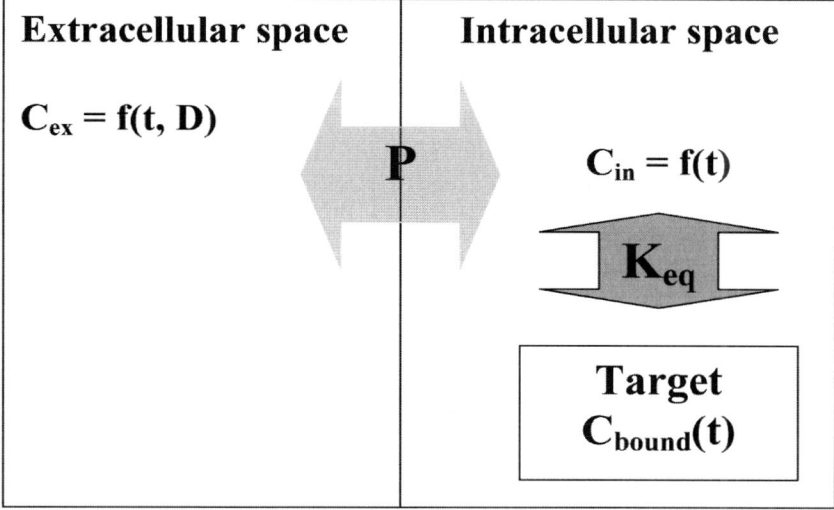

Figure 8-1. A commonly used model of the biological system considered when the molecular structure of drug candidates is designed and tested in vitro (See text for a discussion).

The affinity of a compound to the target protein is indicated by K_{eq}. This number is used in vitro to evaluate a compound's biological activity. From Fig. 8-1 it is clear that biological activity also depends on the intracellular concentration of the drug, C_{in}, which in turn depends on its concentration in extracellular space, C_{ex}, and its ability to cross the plasma membrane as described by the permeability coefficient "P". At present, only these two parameters are used in practice to describe biological systems. It is not surprising therefore that the accuracy of such predictions is only qualitative and for most purposes not accurate enough. Also, it is easy to see that in such a scheme very little information regarding physiology, metabolic processes, and pathophysiology is included. Consequently, there is very little chance of accurately predicting even the quantity of drug associated with the target (cell or, protein) *in vivo*. The reason for this is that the underlying model is too simplistic. A common approach for overcomimg this difficulty is to measure the drug concentration in blood "C_{ex}." In order to interpret such data, pharmacokinetic models are needed. Such models, however, are again drastically simplified. The organism is represented in the form of sets of compartments and drug concentration in each compartment is assumed to be uniform and to depend on the exchange of fluxes between them. The number of compartments is limited by the number of parameters available from experiments (Fig. 8-2). There are two experimental approaches for gathering data in pharmacokinetic studies: the determination of the dependence of drug concentration on time in the blood and the amount eliminated from the organism. Additional information is available from drug concentration in the organs and tissues of sacrificed animals.

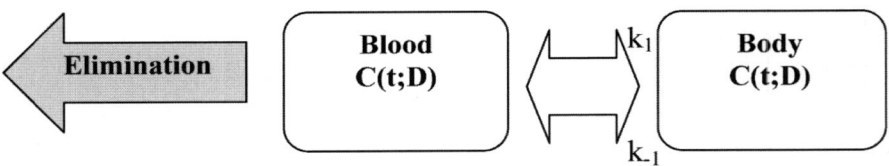

Figure 8-2. A simple example of the two-compartmental pharmacokinetic model.

Other difficulties inherent to this type of approach, despite the obvious simplification, include the financial, man-power, and time expenses needed to provide experimental data of sufficient quality and quantity. This becomes especially troublesome in current research and discovery processes, where a large number of compounds need to be tested. In addition, pahrmacokinetic data exhibit limited correlation to drug potency for curing diseases, since the evaluation of compound's concentration in tissue is a poor indicator of its concentration at the place of action. A series of efforts have been made to devise *in vitro* tests intended to reduce the number of experiments *in vivo*.

The major reason for all these problems is the underlying perception of the drug itself: the belief that a single molecule can meet all the requirements imposed by the task of reaching and selectively altering targeted processes in specific locations without affecting other vital organism functions. It is easy to see that there exist no measures that will quantitate this task and provide guidelines for rational drug design in terms of traditional approaches. Consequently, trial and error still has its place in the modern drug discovery.

Up to this point, the organism has been treated as a black box, which processes active compound. The continuous development of pato-physiology, molecular biology, immunology, genetics, and cell physiology have laid the foundations for innovative approaches in drug design. It has been realized that a large number of parameters decide about compound fate after systemic administration. A number of new approaches and concepts emerged in recent years. First, it was found that certain functions can be ascribed to different elements of a molecule, which will not interfere with the molecule's principal role - biological activity. For this purpose, active compounds have been covalently attached to various synthetic polymers [72, 76, 89, 148, 174] and peptides [20, 113, 154, 171, 175], which were intended to ensure the desired compound properties (solubility) or provide a means of performing a particular function (like enhancing permeation through the endothelium or plasma membrane).

This way of thinking results in the development of sophisticated polymer constructs, which will behave in a predictable way when placed under specific conditions. For example, covalent bonds sensitive to pH are used to induce drug detachment from the polymer when in late endosomes [178]. Such research shows that it is possible to include properties in the structure of a molecule that will ensure its predictable behavior. Manipulations of this kind, however, will provide the drug with only limited specificity.

The design and construction of bloc polymers also showed that there is a shift in thinking about therapeutic agents and, even more importantly, about the organism and the way its processes can be modulated, controlled, and used to the advantage of therapy. The next step was to seek a means of obtaining a higher level of specificity. This can be achieved by attaching the active compound to antibodies with specificity towards the antigens presented by certain cell types [77, 152, 153, 177]. This strategy has already been tasted and successfully employed in a number of new generations of drugs: Rituxan® (IDEC Pharmaceuticals; 1997) Anti-CD20 antibody or relaps/refractory CD-20 positive B-cell against non-Hodgkin's lymphoma and low-grade or follicular-type lymphoma; Herceptin® (Genentech, 1998) it blocks HER2 receptor against HER-2 positive metastatic brest cancer; Mylotarg® (Wyeth Pharmaceuticals, 2000) Anti-CD33 antibody against relapsed/refractory acute myelogenous leukemia; Campath® (Berlax

Laboratories, 2001) with Anti-CD52 antibody against B-cell chronic lymphocytic leukemia; Zevalin® (IDEC Pharmaceuticals, 2002) with Anti-CD20 antibody against Rituximab-failed non-Hodgkins lymphoma; and Iressa® (AstraZeneca, 2003) which blocks epidermal growth factor receptors and tyrosine kinase activity and is used against advanced non-small cell lung cancer. These examples show that when certain information is encoded within construct structures (in this case antibodies towards specific targets), the performance of the resultant formulation may be substantially improved.

There are, however, serious limitations to such an approach. The need to break covalent bonds in order to release the active compound frequently requires enzymatic activity. Also, the drug is not protected before hydrolysis, and limited quantities of drug is delivered in a single "package". To overcome those difficulties, even more complex polymer designs are proposed - for example, triggered drug release from a polymer equipped with antibodies [178]. It was soon realized that much better results can be obtained when such polymers are made capable of aggregating. In that case, the complexity of the resulting formulation drastically increases, allowing a large variety of properties to be encoded within the chemical structure of the components and aggregate topology itself. What is even more important, different components can be mixed into a single structure, effectively extending this possibility even further [23, 90, 105].

In parallel, research on other particulate drug formulations has been intensified. A number of different approaches have been used. The oldest type of formulations were based on spontaneously formed lipid aggregates in the form of liposomes and micelles [64, 107, 109, 111, 117], nanoparticles [19, 58], microspheres [106], microtubules [131], microbubles [179], cochleates [200] and ezymosomes [59]. Each of these formulations has its own specific properties, suitable for certain applications. In the mid nineties, two liposomal preparations have been introduced to the market for systemic applications: Ambisome® - a liposmal formulation of Amphotericin B [1, 18, 187, 192] used for treating systemic fungal infections, and Doxil® - a liposmal formulation of Doxorubicin first used for treating Caposi Sarcoma and later for other indications [36, 66, 83, 110, 169].

The introduction of these particulate formulations has shown that a number of phoarmacokinetic parameters can be manipulated in a range otherwise difficult to obtain. Namely, circulation time has been increased to up to tens of hours, the severity of side effects substantially reduced, and the effects of passive targeting discovered and exploited [23, 130]. Extended persistence time is especially valuable in specialized applications like boron neutron capture therapy, during which active agents are introduced in liposomes [84, 134], or in medical imaging, in which contrast agents last for an extended time and affinity towards pathologically altered tissue is possible [21, 73, 74].

The size of these particulates (from 50 to 200 nm), however, shows that there is yet another aspect of such formulations: interaction with elements of the immunological system [166, 168]. In fact, the immunological system can be stimulated by applying certain types of particulates, even without any active compounds [63, 181-183]. Liposomes are recognized and eliminated by mononuclear cells [161]. This at first seemingly insurmountable difficulty can be used as an advantage in cetrain applications. It makes it possible to target macrophages, therefore opening the door to the development of new strategies for treating intracellular infections [6, 67, 114, 135, 138, 140]. This includes HIV infection [16, 43, 44], where the targets are lymph nodes, in which the infected cells are located [49]. It has also been shown antigens become more efficient and effective vaccines when they are present on the surface of the particulate, including vaccines of liposomal Hepatitis A and B [57, 98], melanoma [204], and HIV [203].

These examples point out the complexity of particulate drug formulations and the variety of their interactions with tissues and cells. It is a challenge to comprehend and quantitate all that information, which opens many possibilies for designing multipurpose and versatile drug delivery devices. It should be stresses at this point that the modern drug is understood as a multi-component system, in which each element has a given function. In the first successful liposomal formulations only limited objectives were targeted. Drug encapsulation ensures its stability, extends its circulation time, and allows for limited targeting. The complexity of particulate drug formulations, however, can be increased practically to infinity. Different properties can be encoded in such structures, so that the cross-talk between the aggregate and biological material can be realized and exploited.

For example, attaching a polymer onto the liposome surface prevents protein association, hence further extending circulation time [3-5]. A variety of polymers have already been tried as surface modifications, including polyethylglycol [26, 75, 112], polysaccharides [8, 91, 93, 194], dextrans [25], and albumins [85], to name a few. The possibility of controlling aggregate persistence in blood provides the time necessary to accumulate in a certain tissue, i.e. tumors or inflammations [136]. Long-lasting liposomes with heamoglobin have even been considered as blood substitutes [115].

Electrostatic charge can be introduced onto the surface, stimulating certain low-specific interactions with selected cells and tissues. For example, a negative charge induces blood coagulation, causing tumor angiogenetic capillaries to become clogged [30, 32], whereas positive charge may induce different types of effects [10, 17]. It has also been found that cationic liposomes are valuable and effective adjuwants in vaccines [137] and that positive charge determines the distribution of liposomes between the vascular and extracascular compartments of tumors [24, 195].

The aggregate surface can be enhanced even further in order to achieve higher specificity by attaching antibodies - these are the so called immunoliposomes [71, 127, 133]. Other special-purpose molecules have been tried as well, including tufsin [2], tissue-specific protein epitopes [22], folate [56, 149, 155, 158], transferin [79], mannose [142], albumin for inducing complement action [159], recombinant tumor necrosis factor-α (160), cationized albumin for enhancing the association of particulates with altered porcine brain microvessel endothelial cells and intact brain capillaries [173], and various other ligands [199].

Such a broad range of chemical signals on the aggregate surface potentially allows a high level of selectivity to be obtained towards certain tissues or even selected cells, drastically reducing non-specific drug accumulation elsewhere. It has already been shown that it is possible to target the endothelium [129] or lymph nodes [139, 143, 144] with high specificity.

There are, however, further possibilities, not directly associated with the aggregate that allow certain properties to be implemented within the aggregate structure and which extend design possibilities. Lipids in aggregates exhibit a high level of cooperativity, manifested in so-called phase transitions (thermotropic and lyotropic), which occur at well-defined temperatures. This, combined with the fact that liposomes become leaky at these temperatures, allows for controlled drug release in artificially heated areas. This results in the design of so-called thermoliposomes, which circulate without releasing drug until a region of elevated temperature is encountered [54, 55, 101, 102]. This approach has an intrinsic problem - the targetted tissue must be first identified and localized. Another approach is to use pH-sensitive liposomes, whose lipid bilayer becomes unstable below certain pH values. This allows for releasing the active compound only after aggregate internalization by cells via the endosomal pathway [15, 60, 61, 68, 69].

These selected examples show the potential of manipulating aggregate properties in order to obtain a customized structure for a particular task. There is, however, an inherent difficulty in the design stage. The fate of the particulate drug formulation inside the body depends on a variety of processes that are different at various locations [12, 164]. In addition to that, aggregate properties should change in response to its actual location. The active compound should stay within the aggregate while in circulation, and released with a certain efficiency when found in the treated tissues. In order to reach this level of predictability, the design process should include not only the aggregate itself, but also each of the body elements that will be encountered on its way to targeted location.

In order to discuss the issues that need to be resolved, let us consider the design of an ideal targeted drug carrier for treating solid tumors. Ideally, after systemic application,

1)   the aggregate should remain stable and undetected by the immuno-logical system, allowing for distribution throughout the body and ensuring the protection of the active compound before hydrolysis and elimination from circulation; when this is successfully accomplished, one can expect that side effects will be reduced,

2)   the aggregate should find the targeted tumor and accumulate there in sufficient quantities, l ensuring high local drug concentration, and therefore higher efficiency and a reduction of the chance of developing drug resistance,

3)   the active compound should be released in order to inflict the desired action, which means that the aggregate should disintegrate and release the drug or penetrate the cell plasma membrane and release its cargo inside the cytoplasm.

At present, there are a number of approaches that address all of the issues listed above. In fact, all of them have been already tested separately or in selected combinations as described before. For example, circulation time can be extended by attaching a polymer to the aggregate surface. Specificity can be achieved by properly selecting the sizes and/or by attaching antibodies to the surface. Drug release can be achieved by temperature or pH triggers. In order to accomplish everything, however, all these properties must be included in the composition and structure of the aggregate. This is a difficult task, since some modifications may contradict others. For example, a surface-bound polymer will protect the aggregate andt at the same time stabilize its structure, making a triggered release via lyotropic phase transition difficult or impossible. There is yet another issue that needs to be addressed in order to design a successful particulate drug formulation. The currently available pharmacokinetic models and approaches are not sufficient to analyze data from experiments *in vivo*. For example, determining the amount of drug in the blood will not provide information on its effective concentration, since the sample will now contain both free drug and drug still remaining in aggregates.

A good example of how difficult it is to predict the outcome of applying particulate drug formulations is the history of thermoliposomes. Thermo-liposomes have been considered as a viable pharmacological approach and it was believed that synchronized rapid drug release from liposomes accumulated in a tumor will improve therapy due to an elevated local concentration. This intuitive is based on reasoning akin to the compartmental pharmacological model. When constructing such liposomes, it was assumed that rapid drug release upon a thermal trigger will result in the desired spatio-temporal concentration profile [54, 55]. Exact numerical simulations, however, shows that the application of thermolyposomes does not improve classical dosage protocols and at best they reduces side effects [51]. There is a simple explanation for this: when the drug is released from liposomes, its local concentration is transiently elevated, but crossing the plasma membrane of cells is a much slower process than diffusion back to the blood system and the drug is readily washed out of the tumor before it has a chance to reach the interior of the targeted cells.

This example illustrates that when a complex drug delivery system is employed, a detailed and careful analysis of all the available data is needed to optimize the formulation itself as well as the protocol of its application. What is even more important, the pathophysiology of the targeted tissue should be implemented into the analysis as well as the physicochemistry of liposomes and drug itself. A temporal and spatial dependence of transport processes needs to be thoroughly analyzed in order to take the advantage of the complexity of the delivery system. For example, the pharmakokinetics and distribution of encapsulated and free drug can be quite different. The reason for this emanates from the difference in the diffusion efficiency or affinity toward cells and/or extracellular structures. When the application of high-affinity antibodies improves carrier selectivity, it will at the same time, reduce their mobility in the volume of the tumor. These and other issues are of particular importance when the pharmacological approach chosen is even more demanding - like that of gene therapy. In the case of gene therapy, the approach used routinely for small-molecule drugs are in fact not applicable. The construction of pharmacokinetic models that will contribute to the new formulation design process should take into account all factors that influence the effectiveness of the therapy. To achieve this, the following points should be taken under consideration.

1) The pharmacokinetics of the free drug form should be predetermined in order to asses the therapeutic concentrations and toxicity limits. This can be done with well-established protocols [145, 198].

2) Special requirements exist for each type of delivery route - particulate formulations open new, previously unavailable routes of delivery. An excellent example is that of the development of transferosomes, which are capable of delivering active compounds into circulation through the skin despite the formidable barrier of the stratum corneum [14, 27]. There are numerous interesting possibilities of applying particulate drug formulations for administration via oral [8, 28, 52], ocular [70, 185], intra-nasal [86-88], and lung [125, 147, 156, 157] routes. Each route has its own peculiarities

and requires specific problems to be solved, such as overcoming the mucous barrier [97, 150].

3) The distribution of the particulate drug formulation within the body should be considered, including factors that affect its stability and mobility if possible, as well as active elimination processes [33, 35, 123]. Tthe effect of aggregate surface modifications on its interaction with morphological blood elements and the endothelium are also important [41, 65]. Enhanced drug carrier association with blood vessel walls is used to treat certain circulation disorders [116, 118, 163]. The effect of serum on aggregate size, distribution, and properties [124], and the stability of the aggregate composition are important issues since supramolecular drug carriers are stabilized only by weak interactions and hydrophobic effects, and the exchange of certain components with surrounding structures is possible or unavoidable [3, 6, 141]. Therefore, the properties of an aggregate will change with the amount time of it resides in the blood. The behavior of drug carriers in circulation seems to be one of the major factors determining the fate of the carrier and subsequently the drug also [31].

4) The assumption of compartmental phormacokinetic models is obviously not sufficient to predict specific carrier accumulation in the targeted tissue and its subsequent spacial distribution, as exemplified by the case of thermoliposomes. Only when inflammation is considered, for which there are no structural barriers for the carrier and drug itself, this assumption seems to be valid [37, 38, 149, 164]. The situation is drastically different when solid tumors are considered. In this case, the tissue volume is highly non-uniform with respect to all relevant properties [47, 78]. Angiogenetic vessels are badly formed, the intercellular space has large necrotic regions and cells are non-uniform and genetically unstable, making them capable of developing drug resistance. All this causes the distribution pattern of the encapsulated and free drug to be complicated [81, 82]. The intratumor space has properties and composition different from the remaining tissues, which can be and is used for selective targeting [46, 50, 79, 99, 100, 122].

5) In many cases the drug's target is located inside the treated cells (in the cytoplasm), therefore overcoming the plasma membrane barrier is required. This is not a simple issue, because if the compound is intended to be released at a targeted tissue and then cross the cell plasma membrane by passive transport, its properties need to be carefully balanced due to the following reasons. The ability to penetrate lipid bilayers implies poor retention in the carrier and therefore large drug loses during circulation. It should be remembered that penetrating the cell plasma membrane is a very slow process, therefore elevated drug concentration should be sustained for a sufficient time, which is not an easy task to achieve. This means that the compound should be hydrophobic, requiring in turn that the carrier fuse with

M. Langner & A. Kozubek

the plasma membrane. Membrane fusion is not a spontaneous process, therefore special features of the carrier are required to perform that task [48, 171, 188, 202]. Either way, special attention should be paid to this step of the delivery process in order to obtain optimal result, otherwise the whole undertaking will fail.

All, the issues mentioned above need to be assessed so that the particulate drug formulation will have the desired properties, orchestrated in a sequence of events that result in the delivery of the compound to its destination. New experimental and theoretical methods have recently been presented that add to the designing process. The concept of combinatorial particulate development, as employed to forming custom immunoliposomes [80] is one of these attempts. This approach obviously simplifies the production processes and design stage.

In all the formulations presented in literature, the design and development processes are badly organized, making the whole undertaking costly and inefficient. There is a need for developing discovery pathways that will allow key parameters to be assessed at early stages and in an orderly meaner, similarly to traditional pahrmaceuticals. In fact, despite the complexity of particulate formulations, the development process can be made much simpler and less time consuming, given an appropriate underlying philosophy of the approach. Aggregates can be ensembled from well-characterized compounds whose properties are predetermined, and only the whole structure and the new active compound need enter the final stages of the development process. Therefore, the costly and time-consuming combinatorial approaches can be reduced or in some cases even eliminated. In order to achieve this, the design and development process need to be organized in a similar manner as that of traditional pharmaceuticals [95, 96]. Organizing the particulate development process should be aided by libraries of components, their aggregates, formulations, and protocols in a manner similar to that of QSAR (Quantitative Structure Activity Relationship) and QSPR (Quantitative Structure Property Relationship) - approaches already used for drugs of low molecular weight. Highly-efficient in vitro methodologies should be developed in order to improve the quality and predictability of new formulations in predetermined situations [119, 151, 167].

2.   SUMMARY AND PERSPECTIVES

Particulate drug formulations are coming of age and beginning to be a viable pharmacological approach. A thorough understanding of physiology, molecular biology, biophysics, and cell physiology is required in order to attempt to search for new generation drug carriers. The main objective of applying particulate drug formulations is to deliver an active compound to a predetermined location, simultaneously eliminating its nonspecific distribution in other organs. It is not possible to reach this objective without monitoring the spatio-temporal distributions of the carrier and/or active compound within the body. This type of data is now available thanks to new developments in imaging techniques, where particulate formulations are already well established as contrast enhancers [11, 21, 73, 74, 103]. Methods such as real-time monitoring of pharmacokinetics and biodistribution [170], multi-photon imaging [176], the ability to monitor intracellular fate of aggregates *in situ* [189], application of High Throughout Screening methodology to pharmacokinetics [190], and development of fast and efficient methods providing simultaneous data on pharmacokinetics and metabolism [191] have created the conditions for obtaining data that are qualitatively different from those available previously. All these technological advances are accompanied by the development of new non-compartmental modeling methods and the implementation of advanced theoretical tools for constructing physiological and metabolic models [92, 94, 128, 184, 186, 193, 205].

Research on targeted drug delivery systems is further accelerated by the emergence of new therapeutic strategies based on proteins [8, 175, 196, 202, 203] and genetic material [40, 108, 132, 197].

Most important of all, these new developments force changes in the perception of therapeutic agents, the organism, tissue, and cells, in turn allowing for a better formulation of pharmacological objectives and providing directions for the research of new strategies and unconventional solutions [12, 13, 29, 146, 165, 201]. Restating some old problems and the emergence of new possibilities are an additional positive outcome of the research on targeted drug delivery systems. Improvements in photodynamic therapy [42], the utilization of the LDL pathway for pharmacological purposes [7, 34], and exploring the possibility of therapeutically applying highly hydrophobic compounds in an effort to interfere in lipid distribution [39, 104, 162] are only a few selected examples of emerging new pharmacological ideas.

3.   REFERENCES

1. Adler-Moore J, Proffitt RT. 2002. AmBisome: liposomal formulation, structure, mechanism of action and pre-clinical experience. J. Antimicrob. Chemother. 49: 21-30.

2. Agrawal AK, Gupta CM. 2000. Tufsin-bearing liposomes in treatment of macrophage-based infections. Adv. Drug Deliv. Rev. 41: 135-46.

3. Ahl PL, Bhata SK, Meers SK, Roberts P, Stevens R, *et al.*, 1997. Enhancement of the *in vivo* circulation lifetime of L-a-distearoylphosphatidylcholine liposomes: importance

of liposomal aggregation versus complement opsonization. Biochim. Biophys. Acta 1239: 370-82.

4. Allen TM. 1998. Oncologic agents in sterically stabilized liposomes: basic considerations. In Long-circulating liposomes: old drugs, new therapeutics., ed. MC Woodle, G Storm, pp. 19-28. Austin TX USA: Landes Bioscience.

5. Allen TM, Menezes DLd, Hansen CB, Moase EH. 1998. Stealth liposomes for the targeting of drug in cancer therapy. Targeting Drugs 6: 61-75.

6. Alving NM, Wassef CR. 1993. Complement-dependent phagocytosis of liposomes. Chem Phys Lipids 64: 239-46.

7. Amin K, Wasan KM, Albercht RM, Heath TD. 2002. Cell association of liposomes with high fluid anionic phospholipid content is mediated specifically by LDL and its receptor, LDLr. J. Pharmaceut. Sci. 91: 1233-44.

8. Aoki H, Fujita M, Sun C, Fuji K, Miyajima K. 1997. High-efficiency entrapment of superoxide dismutase into cationic liposomes containing synthetic aminoglycolipid. Chem Pharm Bull 45: 1327-31.

9. Avdeef A, Testa B. 2002. Physicochemical profiling in drug research: a brief survey of the state-of-the-art of experimental techniques. Cell. Mol. Life Sci. 59: 1681-9.

10. Baczynska D, Widerak K, Ugorski M, Langner M. 2001. Surface charge and the association of liposomes with colon carcinoma cells. Z. Naturforsch. 56c: 872-7.

11. Bakan DA, Weichert JP, Longino MA, Counsell RE. 2000. Polyiodinated triglyceride lipid emulsions for use as hepatoselective contrast agents in CT. Investigative Radiol. 35: 158-69.

12. Bally MB, Lim H, Cullis PR, Mayer LD. 1998. Controling the drug delivery attributes of lipid-based drug formulations. J Liposome Res 8: 299-335.

13. Barenholz Y. 2001. Liposome application: problems and prospects. Curr. Opin. Coll. Interface Sci. 6: 66-77.

14. Barry BW. 2002. Drug delivery routes in skin: a novel approach. Adv. Drug Deliv. Rev. 54: S31-S40.

15. Behr JP. 1997. The proton sponge: a trick to enter cells the viruses did not think of. Chimia 51: 27-30.

16. Bestman-Smith J, Desormeaux A, Tremblay MJ, Bergeron MG. 2000. Targeting cell-free HIV and virally-infected cells with anti-HLA-DR immunoliposomes containing amphotericin B. AIDS 14: 2457-65.

17. Blau S, Jubeh TT, Haupt SM, Rubinstein A. 2000. Drug targeting by surface cationization. Crit. Rev. Therapeut. Drug Carrier Systems 17: 425-65.

18. Boswell GW, Buell D, Bekersky I. 1998. AmBisome (liposomal Amphotericin B): a comparative review. J Clinical Pharmacol 38: 583-92.

19. Brannon-Peppas L, Blanchette JO. 2004. Nanoparticles and targeted systems for cancer therapy. Adv. Drug Deliv. Rev. 56: 1649-59.

20. Brokx RD, Bisland SK, Gariepy J. 2002. Designing peptide scoffolds as drug delivery vehicles. J. Controlled Release 78: 115-23.

21. Brouwers AH, de-Jong DJ, Dams ETM, Oyen WJG, Boerman OC, *et al.*, 2000. Tc-99m-PEG-liposomes for the evaluation of colitis in crohn's disease. J. Drug Targeting 8: 225-33.

22. Brown KC. 2000. New approaches for cell-specific targeting: identification of cell selective peptides from combinatorial libraries. Curr. Opin. Chem. Biol. 4: 16-21.

23. Cammas S, Kataoka K. 1996. Site specific drug-carriers: polymeric micelles as high potental vehicles for biologically active molecules. In Solvents and self-organization of polymers, ed. SE Webber, pp. 83-113. Amsterdam: Kluwer Academic Publishers.

24. Campbell RB, Fukumura D, Brown EB, Mazzola LM, Izumi Y, *et al.*, 2002. Cationic charge determines the distribution of liposomes between the vascular and extracascular compartments of tumors. Cancer Res. 62: 6831-6.

25. Cansell M, Parisel C, Jozefonicz J, Letourneur D. 1999. Liposomes coated with chemically modified dextran interact with human endothelial cells. J Biomed Mater Res 44: 140-8.

26. Cef B, Winterhalter M, Frederik PM, Vallner JJ, Lasic DD. 1997. Stealth liposomes: from theory to product. Adv. Drug Deliv Rev 24: 165-211.

27. Cevc G, Schatzlein A, Richardsen H. 2002. Ultradeformable lipid vesicles can penetrate the skin and other semi-permeable berriers unfragmented. Evidence from double label CLSM experiments and direct size measurements. Biochim. Biophys. Acta 1564: 21-30.

28. Charman WN. 2000. Lipids, lipophilic drugs, and oral drug delivery - some emerging concepts. J. Pharmaceut. Sci. 89: 967-78.

29. Charrois GJR, Allen TM. 2003. Rate of biodistribution of STEALTH liposomes to tumor ad skin: influence of liposome diameter and implications for toxicity and therapeutic activity. Biochim. Biophys. Acta 1609: 102-8.

30. Chiu GNC, Bally MB, Mayer LD. 2001. Selective protein interactions with phosphatidylserine containing liposomes alter the steric stabilization properties of poly(ethylene glycol). Biochim. Biophys. ACta 1510: 56-69.

31. Chiu GNC, Bally MB, Mayer LD. 2002. Effects of phosphatidylserine on membrane incorporation and surface protection properties of exchangeable poly(ethyloene glycol)-conjugated lipids. Biochim. Biophys. Acta 1560: 37-50.

32. Chiu GNC, Bally MB, Mayer LD. 2003. Targeting of antibody conjugated, phsophatidylserine-containing liposomes to vascular cell adhesion molecule 1 for controlled thrombogeneis. Biochim. Biophys. Acta 1613: 115-21.

33. Chow F, Benincosa LJ, Sheth SB, Wilson D, Davis CB, et al. 2002. Pharmacokinetic and pharmacodynamic modeling of humanized anti-factor IX antibody (SB 249417) in humans. Clin. Pharmacol. Ther. 71: 235-45.

34. Chung NS, Wasan KM. 2004. Potential role of the low-density lipoprotein receptor family as mediators of cellular drug uptake. Adv. Drug Deliv. Rev. 56: 1315-34.

35. Clark DE. 2003. In silico prediction of blood-brain barrier permeation. DDT 8: 927-33.

36. Colbern GT, Hiller AJ, Musterer RS, Pegg E, Henderson IC, Working PK. 1999. Significant increase in antitumor potency of doxorubicin HCl by its encapsulation in pegylated liposomes. J. Liposome Res. 9: 523-38.

37. Corvo ML, Boerman OC, Oyen WJG, Bloois LV, Cruz MEM, *et al.*, 1999. Intravenous administration of superoxide dismutase entrapped in long circulating liposomes. II. *In vivo* fate in a rat model of adjuvant arthritis. Biochem. Biophys. Acta 1419: 325-34.

38. Crommelin DJA, van-Resen AJML, Wauben MHM, Storm G. 1999. Liposomes in autoimmune diseases: selected applications in immunotherapy and inflammation detection. J. Controlled Release 62: 245-51.

39. Davidsen J, Jorgensen K, Andersen TL, Mouritsen OG. 2003. Secreted phospholipase A2 as a new enzymatioc trigger nechanism for localised liposomal drug release and absorption in diseased tissue. Biochi. Biophys. Acta 1609: 95-101.

40. Delepine P, Guillaume C, Floch V, Loisel S, Yaouanc JJ, *et al.*, 2000. Cationic phosphonolipids as nonviral vectors: in vitro and in vivo applications. J. Pharmaceut. Sci. 89: 629-38.

41. DelRaso NJ, Foy BD, Gearhart JM, Frazier JM. 2003. Cadmium uptake kinetics in rat hepatocytes: correction for albumin binding. Toxicol. Sci. 72: 19-30.

42. Derycke ASL, de-Witte PAM. 2004. Liposomes for photodynamic therapy. Adv. Drug Delivery Rev. 56: 17-30.

43. Desormeaux A, Bergeron MG. 1995. Targeting HIV with liposome-encapsulated antivirals. Zbl Bakt 282: 225-31.

44. Desormeaux A, Bergeron MG. 1998. Liposomes as drug delivery systems: a strategic approach for the treatment of HIV infection. J Drug Tarheting 6: 1-15.

45. Devis SS, Illum L. 1998. Drug delivery systems for challenging molecules. Int. J. Pharmaceutics 176: 1-8.

46. DiTizio V, Karlgard C, Lilge L, Khoury AE, Mittelman MW, DiCosmo F. 2000. Localized drug delivery using crosslinked gelatin gels containing liposomes: factors influencing liposome stability and drug release. J. Biomed. Mater. Res. 51: 96-106.

47. Drumond DC, Meyer O, Hong K, Kirpotin DB, Papahadjopoulos D. 1999. Optimizing liposomes for delivery of chemotherapeutic agents to solid tumors. Pharmacological Rev. 51: 691-743.

48. Duzgunes N, Nir S. 1999. Mechanisms and kinetics of liposome-cell interactions. Adv Drug Deliv Rev 40: 3-18.

49. Duzgunes N, Pretzer E, Simoes S, Slepushkin V, Konopka K, *et al.*, 1999. Liposome-mediated delivery of antiviral agents to human immunodeficiency virus-infected cells. Mol Mem Biol 16: 111-8.

50. Ehrhardt C, Kneuer C, Bakowsky U. 2004. Selectin - an emerging target for drug delivery. Adv. Drug. Deliv. Rev. 56: 527-49.

51. El-Kareh AW, W. Secomb T. 2000. A Mathematical Model for Comparison of Bolus Injection,Continuous Infusion, and Liposomal Delivery of Doxorubicin to Tumor Cells. Neoplasia 2: 325-38.

52. Foster N, Hirst BH. 2005. Exploiting receptor biology for oral vaccination with biodegradable particulates. Adv. Drug Deliv. Rev. 57: 431-50.

53. Fouchecourt MO, Beliveau M, Krishnan K. 2001. Quantitative structure-pharmacokinetic relationship modeling. Sci. Total Environment 274: 125-35.

54. Gaber MH, Hong K, Huang SK, Papahadjopoulos D. 1995. Thermosensitive sterically stabilized liposomes: formulation and *in vitro* studies on mechanism of doxorubicin release by bovine serum and human plasma. Pharmaceut Res 12: 1407-16.

55. Gaber MH, Wu NZ, Hong K, Huang SK, Dewrist MW, Papahadjopoulos D. 1996. Thermosensitive liposomes: extravasation and release of contents in tumor microvascular networks. Int J Radiation Oncol Biol Phys 36: 1177-87.

56. Gabizon A, Shmeeda H, Horowitz AT, Zalipsky S. 2004. Tumor cell targeting of liposome-entrapped drugs with phospholipid-anchored folic acid-PEG conjugates. Adv. Drug Deliv. Rev. 56: 1177-92.

57. Garcia M, Alsina MA, Reig F, Haro I. 2000. Liposomes as vehicles for the presentation of a synthetic peptide containing an epitope of hepatitis A virus. Vaccine 18: 276-83.

58. Garcia-Fuentes M, Torres D, Alonso MJ. 2002. Design of lipid nanoparticles for the oral delivery of hydrophobic macromolecules. Coll. Surf. B: Biointerfaces 27: 159-68.

59. Gaspar MM, Martins MB, Corvo ML, Cruz MEM. 2003. Design and characterization of enzymosomes with surface-exposed superoxide dismutase. Biochim. Biophys. Acta 1609: 211-7.

60. General S, Thunemann AF. 2001. pH-sensitive nanoparticles of poly(amino acid) dodecanoate complexes. Interant. J. Pharmaceut. 230: 11-24.

61. Gerasimov OV, Boomer JA, Qualls MM, Thompson DH. 1999. Cytosolic drug delivery using pH- and light-sensitive liposomes. Adv Drug Delivery Rev 38: 317-38.

62. Gershanik T, Haltner E, Lehr CM, Benita S. 2000. Charge-dependent interaction of self-emulsifying oil formulations with Caco-2 cells monolayers: binding, effects on barrier function and cytotoxicity. Intl. J. Pharmaceut. 211: 29-36.

63. Gould-Fogerite S, Kheiri MT, Zhang F, Wang Z, Scolpino AJ, *et al.*, 1998. Targeting immune response induction with cocheate and liposome-based vaccines. Adv Drug Delivery Rev 32: 273-87.

64. Gregoriadis G, McCormack B, Morrison Y, Saffie R, Zadi B. 1998. Liposomes in drug targeting. Cell Biol.: a laboratory handbook 4: 131-40.

65. Gulden M, Morchel S, Seibert H. 2003. Serum albumin binding at cytotoxic concentrations of chemicals as determined with a cell proliferation assay. Toxicology Lett. 137: 159-68.

66. Guo LSS. 2001. Amphotericin B colloidal dispersion: an improved antifungal therapy. Adv. Drug Deliv. Rev. 47: 149-63.

67. Gursoy A. 2000. Liposome-encapsulated antibiotics: physicochemical and antibacterial properties, a review. S.T.P. Pharma. Sci. 10: 285-91.

68. Hafez IM, Ansell S, Cullis PR. 2000. Tunable pH-sensitive liposomes composed of mixtures of cationic and anionic lipids. Biophys. J. 79: 1438-46.

69. Hafez IM, Cullis PR. 2001. Role of lipid polymorphism in intracellular delivery. Adv. Drug Deliv. Rev. 47: 139-48.

70. Han YH, Sweet DH, Hu DN, Pritchard JB. 2001. Characterization of a novel cationic drug transporter in human retinal pigment epithelial cells. J. Pharmacol. Exp. Therapeu. 296: 450-7.

71. Hansen CB, Kao GY, Moase EH, Zalipsky S, Allen TM. 1995. Attachment of antibodies to sterically stabilized liposomes: evaluation, comparision and optimization of coupling procedures. Biochi Biophys Acta 856: 556-81.

72. Harada A, Kataoka K. 1998. Novel polyion complex micelles entrapping enzyme molecules in the core: preparation of narrowly-distributed micelles from lysozyme and poly(ethylene glycol)-poly(aspartic acid) block copolymer in aqueous medium. Macromolecules 31: 288-94.

73. Harrington KJ, Rowlinson-Busza G, Syrigos KN, Abra RM, Uster PS, *et al.*, 2000. Influence of tumor size on uptake of 111In-DTPA-labelled pegylated liposomes in a human tumour xenograft model. Brit. J. Cancer 83: 684-8.

74. Harrington KJ, Rowlinson-Busza G, Syrigos KN, Uster PS, Abra RM, Stewart JSW. 2000. Biodistribution and pharmacokinetics of 111In-DTPA-labelled pegylated liposomes in a human tumour xenograft model: implications for novel targeting strategies. Brit. J. Cancer 83: 232-8.

75. Harrington KJ, Rowlinson-Busza G, Syrigos KN, Uster PS, Vile RG, Stewart JSW. 2000. Pegylated liposomes have potential as vehicles for intratumoral and subcutaneous drug delivery. Clinical Cancer Res. 6: 2528-37.

76. Hayashi H, Kono K, Takagishi T. 1999. Temperature sensitization of liposomes using copolymers of N-isopropylacrylamide. Bioconjugate Chem 10: 412-8.

77. Holt LJ, Herring C, Jespers LS, Woolven BP, Tomlinson IM. 2003. Domain antibodies: proteins for therapy. Trends Biotechnol. 21: 484-90.

78. Houshamand P, Zlotnik A. 2003. Targeting tumor cells. Curr. Opin. Cell Biol. 15: 640-4.

79. Ishida O, Maruyama K, Tanahashi H, Iwatsuru M, Sasaki K, *et al.*, 2001. Liposome bearing polyethyleneglycol-coupled transferrin with intracellular targeting property to the solid tumors *in vivo*. Pharmaceut. Res. 18: 1042-8.

80. Ishida T, Iden DL, Allen TM. 1999. A combinatorial approach to producing sterically stabilized (stealth) immunoliposomal drugs. FEBS Lett. 460: 129-33.

81. Jain RK. 1999. Transport of molecules, particles, and cells in solid tumors. Annu. Rev. Biomed. Eng. 1: 241-63.

82. Jain RK. 2001. Delivery of molecular and cellular medicine to solid tumors. Adv. Drug Deliv. Rev. 46: 149-68.

83. Janknegt R, vanEtten EWM, deMarie S. 1996. Lipid formulations of amphotericin B. Curr Opin Infect Dis 9: 403-6.

84. Johnsson M, Bergstrand N, Edwards K. 1999. Optimization of drug loading procedures and characterization of liposomal formulations of two novel agents intended for boron neutron capture therapy (BNCT). J Liposome Res 9: 53-79.

85. Johnstone SA, Masin D, Mayer L, Bally MB. 2001. Surface-associated serum proteins inhibit the uptake of phosphatidylserine and poly(ethylene glycol) liposomes by mouse macrophages. Biochim. Biophys. Acta 1513: 25-37.

86. Jung BH, Chung BC, Chung S, Lee M, Shim C. 2000. Prolong delivery of nicotine in rats via nasal administration of proliposomes. J. Contralled Release 66: 73-9.

87. Jung BH, Chung BC, Chung S, Shim C. 2001. Different pharmacokinetics of nicotine followig intravenous administration of nicotine base and nicotine hydrogen tartarate in rats. J. Contralled Release 77: 183-90.

88. Jung BH, Chung SJ, Shim CK. 2002. Proliposomes as prolonged intranasal drug delivery systems. STP Pharma Sci. 1: 33-8.

89. Kataoka K. 1997. Targetable polymeric drugs. In Controlled drug delivery: challenges and strategies., ed. K Park, pp. 49-71. Washington D.C.: American Chemical Society.

90. Kataoka K, Harada A, Nagasaki Y. 2001. Block copolymer micelles for drug delivery: design, characterization and biological significance. Adv. Drug Deliv. Rev. 47: 113-31.

91. Kato E, Taguchi A, Sakashita S, Akiyoshi K, Sunamoto J. 2000. Synthesis and function of sialic acid-conjugated cholesterols as ganglioside analogs: their reconstitution to liposomes and interaction with rat lymphocytes. Proc. Japan Acad. 76 B: 63-7.

92. Kavanagh BD, Secomb TW, Hsu R, Lin P, Venitz J, Dewhirst MW. 2002. A theoretical model for the effects of reduced hemoglobin-oxigen affinity on tumor oxygenation. Int. J. radiation Oncology Biol. Phys. 53: 172-9.

93. Kawakami S, Munakata C, Fumoto S, Yamashita F, Hashida M. 2000. Targeted delivery of prostaglandin E1 to hepatocytes using galactosylated liposomes. J. Drug Targeting 8: 137-42.

94. Kennedy RR, French RA, Spencer C. 2002. Predictive accuracy of a model of volatile anesthetic uptake. Anesth. Analg. 95: 1616-21.

95. Kerns EH, Di L. 2003. Pharmaceutical profiling in drug discovery. DDT 8: 316-23.

96. Khandurina J, Guttman A. 2002. Microchip-based high-throughput screening analysis of combinatorial libraries. Curr. Opin. Chem. Biol. 6: 359-66.

97. Khanvilkar K, Donovan MD, Flanagan DR. 2001. Drug transfer through mucus. Adv. Drug Deliv. Rev. 48: 173-93.

98. Kim CK, Jeong EJ, Kim MH. 2000. Comparision of *in vivo* fate and immunogenicity of hepatitis B surface antigen incorporated in cationic and neutral liposomes. J. Microencapsulation 17: 297-306.

99. Koivunen E, arap W, Valtanen H, Haininsalo A, Pebate-Medina O, *et al.*, 1999. Cancer therapy with a novel tumor-targeting gelatinase inhibitor selected by phage peptide display. Nat. Biotechnol. 17: 768-74.

100. Koivunen E, Ranta TM, Annila A, Taube S, Uppala A, *et al.*, 2001. Inhibition of beta(2) integrin-mediated leukocyte cell adhesion by leucine-leucine-glycine motif-contaning peptides. J. Cell. Biol. 153: 905-16.

101. Kong G, Anyarambhatla G, Petros WP, Braun RD, Colvin OM, *et al.*, 2000. Efficiency of liposomes and hyperthermia in a human tumor xenograft model: importance of triggered drug release. Cancer Res. 60: 6950-7.

102. Kono K, Nakai R, Morimoto K, Takagishi T. 1999. Thermosensitive polymer-modified liposomes that release contents around physiological temperature. Biochim Biophys Acta 1416: 239-50.

103. Krause W. 1999. Delivery of diagnostic agents in computed tomography. Adv Drug Delivery Rev 37: 159-73.

104. Kurz M, Scriba GKE. 2000. Drug-phospholipid conjugates as potential prodrugs: synthesis, characterization, and degradation by pancreatic phospholipase A2. Chem. Phys. Lipids 107: 143-57.

105. Kwon GS, Okano T. 1996. Polymeric micelles as new drug carriers. Adv. Drug. Delivery. Rev. 21: 107-16.

106. Lacasse FX, Filion MC, Phillips NC, Escher E, McMullen JN, Hilden P. 1998. Influence of surface properties at biodegradable microsphere surfaces: effects on plasma protein adsorption and phagocytosis. Pharm Res 15: 312-7.

107. Langner M. 2000. Effect of liposome molecular composition on its ability to carry drugs. Polish J Pharmacol 52: 3-14.

108. Langner M. 2000. The intracellular fate of non-viral DNA carriers. Cell. Molec. Biol. Lett. 5: 295-313.

109. Langner M, Kral T. 1999. Liposome-based drug delivery systems. Pol J Pharmacol 51: 211-22.

110. Lasic DD. 1996. Doxorubicin in sterically stabilized liposomes. Nature 380: 5611-20.

111. Lasic DD. 1997. Recent developments in medical applications of liposomes: sterically stabilized liposomes in cancer therapy and gene delivery in vivo. J Contr Rel 48: 203-22.

112. Lasic DD, Needham D. 1995. The "Stealth" Liposome: a prototypical biomaterial. Chem Rev 95: 2601-34.

113. Lazar GA, Marshall SA, Plecs JJ, Mayo SL, Desjarlais JR. 2003. Designing proteins for therapeutic applications. Curr. Opin. Struct. Biol. 13: 513-9.

114. Leitzke S, Bucke W, Borner K, Muller R, Hahn H, Ehlers S. 1998. Rationale for and efficacy of prolonged-interval treatment using liposome-encapsulated amikacin in experimental Mycobacterium avium infection. Antimicrob Agents Chemother 42: 459-61.

115. Leslie SB, Puvvada S, Ratna BR, Rudolph AS. 1996. Encapsulation of hemoglobin in a bicontinuous cubic phase lipid. Biochim Biophys Acta 1285: 246-54.

116. Lestini BJ, Sagnella SM, Xu Z, Shive MS, Richter NJ, et al., 2003. Surface modification of liposomes for selective cell targeting in cardiovascular drug delivery. J. Controlled Release 78: 235-47.

117. Lian T, Ho RJY. 2001. Trends and developments in liposome drug delivery systems. J. Pharmaceut. Sci. 90: 667-80.

118. Liaw J, Aoyagi T, Kataoka K, Sakurai Y, Okano T. 1999. Permeation of PEO-PBLA-FITC polymeric micelles in aortic endothelial cells. Pharm Res 16: 213-20.

119. Lohmann C, Huwel S, Galla HJ. 2002. Predicting blood-brain barrier permeability of drugs: evaluation of different in vitro assays. J. Drug Targeting 10: 263-76.

120. Loidl-Stahlhofen A, Eckert A, Hartmann T, Schottner M. 2001. Solid-supported lipid membranes as a tool for determination of membrane affinity: high-throughput screening of a physicochemical parameters. J. Pharmaceut. Sci. 90: 599-606.

121. Loidl-Stahlhofen A, Hartmann T, Schottner M, Rohring C, Brodowsky H, et al., 2001. Multilamellar liposomes and solid-supported lipid membranes (TRANSIL): screening

of lipid-water partitioning toward a high-throughput scale. Pharmaceut. Res. 18: 1782-8.

122. Lopes-de-menezes DE, Pilarski LM, Belch AR, Allen TM. 2000. Selective targeting of immunoliposomeal doxorubicin against human multiple myeloma *in vitro* and *ex vivo*. Biochim. Biophys. Acta 1466: 205-20.

123. Mager DE, Jusko WJ. 2002. Quantitative structure-pharmacokinetic/pharmacodynamic relationship of corticosteroids in man. J. Pharmaceut. Sci. 91: 2441-51.

124. Makabi-Panzu B, Gourde P, Desormeaux A, Bergeron MG. 1998. Intracellular and serum stability of liposomal 2',3'-dideoxycytidine. Effect of lipid composition. Cell Mol Biol 44: 277-84.

125. Makino K, Yamamoto N, Higuchi K, Harada N, Ohshima H, Terada H. 2003. Phagocytic uptake of polystyrene microspheres by alveolar macrophages: effect of the size and surface properties of the microspheres. Colloids and SUrface B: Biointerfaces. 27: 33-9.

126. Mannhold R, van-de-Waterbeemd H. 2001. Substructure and whole molecule approaches for calculating logP. J. Computer-Aided Molec. Design 15: 337-54.

127. Marcucci F, Lefoulon F. 2004. Active targeting with particulate drug carriers in tumor therapy: fundamentals and recent progress. DDT 9: 219-28.

128. McGinnity DF, Riley RJ. 2001. Predicting drug pharmacokinetics in human from *in vitro* metabolism studies. Biochem. Soc. Trans. 29: 135-40.

129. McIntosh DP, Tan X, Oh P, Schnitzer JE. 2002. Targeting endothelium and its dynamic caveolae for tissue-specific transcytosis *in vivo*: a pathway to overcome cell barriers to drug and gene delivery. PNAS 99: 1996-2001.

130. Mehta J, Kelsey S, Chu P, Powles R, Hazel D, *et al.*, 1997. Amphotericin B lipid complex (ABLC) for the treatement of confirmed or presume fungal infections in immunocompromised patients with hematologic malignancies. Bone Marrow Transplant 20: 39-43.

131. Meilander NJ, Yu X, Ziats NP, Bellamkonda RV. 2001. Lipid-based microtubular drug delivery vehicles. J. Controlled Release 71: 141-52.

132. Miller AD. 1998. Cationic liposomes for gene therapy. Angew. Chem. Int. Ed. 37: 1768-85.

133. Mills JK, Needham D. 1999. Targeted drug delivery. Exp Opin Ther Patents 9: 1499-513.

134. Miyamoto M, Hirano K, Ichikawa H, Fukumori Y, Akine Y, Tokuuye K. 1999. Preparation of gadolinium-containing emulsions stabilized with phosphatidylcholine-surfactant mixtures for neutron-capture therapy. CHem Pharmaceut Bull 47: 203-8.

135. Moghimi SM, Patel HM. 2002. Modulation of murine liver macrophage clearance of liposomes by diethylstilbestrol. The effect of vesicle surface charge and a role for the complement receptor Mac-1 (CD11b/CD18) of newly recruited macrophages in liposome recognition. J. Controlled Release 78: 55-65.

136. Mosqueira VCF, Legrand P, Morgat J-L, Vert M, Mysiakine E, *et al.*, 2001. Biodistribution of long-circulating PEG-grafted nanocapsules in mice: effect of PEG chain length and density. Pharmaceut. Res. 18: 1411-9.

137. Nakanishi T, Kunisawa J, Hayashi A, Tsutsumi Y, Kubo K, *et al.*, 1999. Positively charged liposome functions as an efficient immunoadjuvant in inducing cell-mediated immune response to soluble proteins. J Controlled Release 61: 233-40.

138. Nightingale SD, Saletin SL, Swenson CE, Lawrence AJ, Watson DA, *et al.*, 1993. Liposome-encapsulated gentamicin treatement of Mycobacterium avium-Mycobacterium intracellulare complex bacteremia in AIDS patients. Antymicrob Agents Chemother 37: 1869-72.

139. Nishioka Y, Yoshino H. 2001. Lymphotic targeting with nanoparticulate system. Adv. Drug Deliv. Rev. 47: 55-64.

140. Oh YK, Nix DE, Straubinger RM. 1995. Formulation and efficacy of liposome-encapsulated antibiotics for therapy of intracellular Mycobacterium avium infection. Antimicrob Agents Chemother 39: 2104-11.

141. Oja CD, Semple SC, Chonn A, Cullis PR. 1996. Influence of dose on liposome clearance: critical role of blood proteins. Biochim Biophys ACta 1281: 31-7.

142. Opanasopit P, Higuchi Y, Kawakami S, Yamashita F, Nishikawa M, Hashida M. 2001. Involvement of serum mannan binding protein and mannose receptors in uptake of mannosylated liposomes by macrophages. Biochim. Biophys. Acta 1511: 134-45.

143. Oussoren C, Storm G. 1999. Role of macrophages in the localisation of liposomes in lymph nodes after subcutaneous administration. Internatl. J. Pharmaceutics 183: 37-41.

144. Oussoren C, Zuidema J, Crommelin DJA, Storm G. 1997. Lymphatic uptake and biodistribution of liposomes after subcutaneous injection. II. Influence of liposomal size, lipid composition and lipid dose. Biochim Biophys Acta 1328: 261-72.

145. Panetta JC, Yanishevski Y, Pui CH, Sandlund JT, Rubnitz J, *et al.*, 2002. A mathematical model of *in vivo* methotrexate accumulation in acute lymphoblastic leukemia. Cancer Chemother. Pharmacol. 50: 419-28.

146. Papahadjopoulos D. 1996. Fate of liposomes *in vivo*: a brief introductory review. J. Liposome Res 6: 3-17.

147. Parr MJ, Masin D, Cullis PR, Bally MB. 1997. Accumulation of liposomal lipid and encapsulated doxorubicin in Murine Lewis lung carcinoma: the lack of beneficial effects by coating liposomes with poly(ethylene glycol). J Pharm Exp Therapetics 280: 1319-27.

148. Patnum D, Kopecek J. 1995. Polymer conjugates with anticancer activity. Adv Polymer Sci 122: 55-123.

149. Paulos CM, Turk MJ, Breur GJ, Low PS. 2004. Folate receptor-mediated targeting of therapeutic and imaging agents to activated macrophages in rheumatoid arthritis. Adv. Drug Deliv. Rev. 56: 1205-17.

150. Prego C, Garcia M, Torres D, Alonso MJ. 2005. Transmucosal macromolecular drug delivery. J. Controlled Release 101: 151-62.

151. Rausch JM, Wimley WC. 2001. A high-throughput screen for identifying transmembrane pore-forming peptides. Analyt. Biochem. 293: 258-63.

152. Rechlaender BN, Cho MJ. 2001. Antibodies as drug carriers. I. For proteins. Pharmaceut. Res. 18: 753-60.

153. Rechlaender BN, Cho MJ. 2001. Antibodies as drug carriers. II. For small molecules. Pharmaceut. Res. 18: 745-52.

154. Rensen PCN, de-Vrueh RLA, Kuiper J, Bijsterbosch MK, Biessen EAL, van-Berkel TJC. 2001. Recombinant lipoproteins: lipoprotein-like lipid particles for drug targeting. Adv. Drug Deliv. Rev. 47: 251-76.

155. Rui Y, Wang S, Low PS, Thompson DH. 1998. Diplasmenylcholine-folate liposomes: An efficient vehicle for intracellular drug delivery. J Amer Chem Soci 120: 11213-8.

156. Saari M, Vidgren MT, Koskinen MO, Turjanmaa VMH, Niminen MM. 1999. Pulmonary distribution and clearance of two beclomethasone liposome formulations in healthy volunteers. Inter J Pharm 181: 1-9.

157. Sachetelli S, Beaulac C, Riffon R, Lagace J. 1999. Evaluation of the pulmonary and systemic immunogenicity of fluidosomes, a fluid liposomal-tobramycin formulation for the treatment of chronic infections in lungs. Biochim Biophys Acta 1428: 334-40.

158. Saul JM, Annapragada A, Natarajan JV, Bellamkonda RV. 2003. Controlled targeting of liposomal doxorubicin via the folate receptor *in vitro*. J. Controlled Release 92: 49-67.

159. Savay S, Szebeni J, Baranyi L, Alving CR. 2002. Potentiation of liposome-induced complement activation by surface-bound albumin. Biochim. Biophys. Acta 1559: 79-86.

160. Savva M, Duda E, Huang L. 1999. A genetically modified recombinant tumor necrosis factor-a conjugated to the distal terminals of liposomal surface grafted polyethylene-glycol chains. Internal J Pharmaceut 184: 45-51.

161. Scherphof GL, Kamps JAAM. 2001. The role of hepatocytes in the clearance of liposomes from the blood circulation. Progress Lipid Res. 40: 149-66.

162. Senchenkov A, Litvak DA, Cabot MC. 2001. Targeting ceramide metabolism - a strategy for overcoming drug resistance. J. Natl. Cancer Inst. 93: 347-57.

163. Spragg DD, Alford DR, Greferath R, Larsen CE, Lee KD, *et al.*, 1997. Immunotargeting of liposomes to activated vascular endothelial cells: a strategy for site-selective delivery in the cardiovascular system. Proc Natl Acad Sci USA 94: 8795-800.

164. Storm G, Bakker-Woudenberg IAJM, Schiffelers RM, Oyen WJG, Crommelin DJA, *et al.*, 1998. Diagnostic and therapeutic targeting of infections and inflamatary diseases using sterically stabilized liposomes. In Targeting of Drugs 6: Strategies for Stealth Therapeutic System., ed. Gregoriadis, McCormack, pp. 121-30. New York: Plenum Press.

165. Storm G, Crommelin DJA. 1998. Liposomes: quo vadis? PSTT 1: 19-31.

166. Storni T, Kundig TM, Senti G, Johansen P. 2005. Immunity in response to particulate antigen-delivery systems. Adv. Drug Delivery Rev. 57: 333-55.

167. Sugano K, Hamada H, Machida M, Ushio H. 2001. High throughput prediction of oral adsorption: improvement of the composition of the lipid solution used in parallel artificial membrane permeability assay. J. Biomolecular Screening 6: 189-96.

168. Szebeni J, Baranyi L, Savay S, Lutz HU, Jelezarova E, *et al.*, 2000. The role of complement activation in hypersensitivity to pegylated liposomal Doxorubicin (DOXIL). J. Liposome Res. 10: 467-81.

169. Tardi PG, Boman NL, Cullis PR. 1996. Liposomal doxorubicin. J Drug Targeting 4: 129-40.

170. Taylor DL, Woo ES, Giuliano KA. 2001. Real-time molecular and cellular analysis: the new frontier of drug discovery. Curr. Opin. Biotechnol. 12: 75-81.

171. Temsamani J, Vidal P. 2004. The use of cell-penetrating peptides for drug delivery. DDT 23: 1012-9.

172. Testa B. 1997. Drugs as chemical messages: molecular sytructure, biological context, and structure-activity relationship. Med. Chem. Res. 7: 340-65.

173. Thole M, Nobmann S, Huwyler J, Fricker G. 2002. Uptake of cationized albumin coupled liposomes by cultured porcine brain microvessel endothelial cells and intact brain capillaries. J. Drug Targeting 10: 337-44.

174. Torchilin VP. 2001. Structure and design of polymeric surfactant-based drug delivery systems. J. Controlled Release 73: 137-72.

175. Torchilin VP, Lukyanov AN. 2003. Peptide and protein drug delivery to and into tumors: challenges and solutions. Drug Discov. Today 8: 259-66.

176. Tozer GM, Ameer-Beg SM, Baker J, Barber PR, Hill SA, *et al.*, 2005. Intravital imaging of tumor vascular networks using multi-photon fluorescence microscopy. Adv. Drug Deliv. Rev. 57: 135-52.

177. Trail PA, Bianchi AB. 1999. Monoclonal antibody drug conjugates in the treatment of cancer. Curr Opin Immunol 11: 584-8.

178. Ulbrich K, Subr V. 2004. Polymeric anticancer drugs with pH-controlled activation. Adv Drug Deliv Rev 56: 1023-50.

179. Unger EC, Porter T, Culp W, Labell R, Matsunaga T, Zutshi R. 2004. Therapeutic applications of lipid-coated microbubbles. Adv. Drug. Deliv. Rev. 56: 1291-314.

180. van-de-Waterbeemd H, Smith DA, Beaumont K, Walker DK. 2001. Property-based design: optimization of drug absorption and pharmacokinetics. J. Med. Chem. 44: 1313-33.

181. van-Slooten KL, Storm G, Zoephel A, Kupcu Z, Boerman O, *et al.*, 2000. Liposomes containing interferon-gamma as adjuvant in tumor cell vaccines. Pharmaceut. Res. 17: 42-8.

182. van-Slooten ML, Boerman O, Romoren K, Kedar E, Crommelin DJA, Storm G. 2001. Liposomes as sustained release system for human interferon-g biopharmaceutical aspects. Biochim. Biophys. Acta 1530: 134-45.

183. van-Slooten ML, Kircheis R, Koppenhagen FJ, Wagner E, Storm G. 1999. Liposomes as cytokine-supplement in tumor cell-based vaccines. Int. J. Pharmaceutics 183: 33-6.

184. van-Zuylen L, Karlsson MO, Verweij J, Brouwer E, de-Bruijn P, *et al.*, 2001. Pharmacokinetic modeling of paclitaxal encapsulation in cremophor EL micelles. Cancer Chemother. Pharmacol. 47: 309-18.

185. Velpandian T, Gupta SK, Gupta YK, Biswas NR, Agarwal HC. 1999. Ocular drug targeting by liposomes and their corneal interactions. J Microencapsulation 16: 243-50.

186. Veng-Pedersen P. 2001. Noncompartmentally-based pharmacokinetic modeling. Adv. Drug Deliv. Rev. 48: 265-300.

187. Walsh TJ, Hiemenz JW, Seibel NL, Perfect JR, Horwith G, *et al.*, 1998. Amphotericin B lipid complex for invasive fungal infections: analysis of safety and efficacy in 556 cases. Clinic Infect Disease 26: 1383-96.

188. Watabe A, Yamaguchi T, Kawanishi T, Uchida E, Eguchi A, *et al.*, 1999. Target-cell specificity of fusogenic liposomes: membrane fusion-mediated macromolecule delivery into human blood mononuclear cells. Biochim Biophys Acta 1416: 339-48.

189. Watson P, Jones AT, Stephens DJ. 2005. Intracellular trafficking pathways and drug delivery: fluorescence imaging of living and fixed cells. Adv. Drug Deliv. Rev. 57: 43-61.

190. Watt AP, Morrison D, Evans DC. 2000. Approaches to higher-throughput pharmacokinetics (HTPK) in drug discovery. DDT 5: 17-24.

191. White RE. 2000. High-throughput screening in drug metabolism and pharmacokinetic support of drug discovery. Annu. Rev. Pharmacol. Taxicol. 40: 133-57.

192. Working PK. 1999. Amphotericin B colloidal dispersion. Chemotherapy 45: 15-26.

193. Wright JG, Boddy AV. 2001. All half-lives are wrong, but some half-lifes are useful. Clin. Pharmacokinet. 40: 237-44.

194. Yanagihara K, Kato E, Hitomi S, Sunamoto J, Wada H. 1999. Activation of human T lymphocytes by ganglioside-containing liposomes. Glycoconjugate J 16: 59-65.

195. Yang SC, Benita S. 2000. Enhanced adsorption and drug targeting by positively charged submicron emulsions. Deug Dev. Res. 50: 476-86.

196. Ye Q, Asherman J, Stevenson M, Brownson E, Katre NV. 2000. DepoFoamTM technology: a vehicle for controlled delivery of protein and peptide drugs. J. Controlled Release 64: 155-66.

197. You J, Kamihira M, Iijima S. 1998. Enhancement of transfection efficiency using ligand-modified lipid vesicles. J Ferment Bioengineer 5: 525-8.

198. Yu H, Adedoyin A. 2003. ADME-Tox in drug discovery integration of experimental and computational technologies. Drug Disc. Today 8: 852-61.

199. Zalipsky S, Gittelman J, Mullah N, Qazen MM, Harding JA. 1998. Biologically active ligand-bearing polymer-grafted liposomes. In Strategies for Stealth Therapeutic Systems., ed. Ga McCormack. New York: Plenum Press.

200. Zarif L, Graybill JR, Perlin D, Mannino RJ. 2000. Cochleates: new lipid-based drug delivery systems. J. Liposome Res. 10: 523-38.

201. Zasadzinski JA, Kisak E, Evans C. 2001. Complex vesicle-based structures. Curr. Opin. Coll. Interface Sci. 6: 85-90.

202. Zelphati O, Wang Y, Kitada S, Reed JC, Felgner PL, Corbeil J. 2001. Intracellular delivery of proteins with a new lipid-mediated delivery system. J. Biol. Chem. 276: 35103-10.
203. Zheng L, Huang XL, Fan Z, Borowski L, Wilson CC, Rinaldo CR. 1999. Delivery of liposome-encapsulated HIV type 1 proteins to human dendritic cells for stimulation of HIV type 1-specific memory cytotoxic T lymphocyte tesponses. AIDS Res Hum Retrovir 15: 1011-20.
204. Zhou WZ, Kaneda Y, Huang SKS, Mrishita R, Hoon DSB. 1999. Protective immunization against melanoma by gp100 DNA-HVJ-liposome vaccine. Gene Therapy 6: 1768-73.
205. Zuideveld KP, van-Gestel A, Peletier LA, van-der-Graff PH, Danhof M. 2002. Pharmacokinetic-pharmacodynamic modelling of the hypothermic and corticosterone effects of the 5-HT1A receptor agonist flesinocan. Eur. J. Pharmacol. 445: 43-54.

Chapter 9

SYNTHETIC VECTORS FOR GENETIC DRUG DELIVERY

Paulina Wyrozumska[1,2], Katarzyna Stebelska[2], Michal Grzybek[1] and Aleksander F. Sikorski[1,3]

[1]University of Wroclaw, Institute of Biochemistry and Molecular Biology, [2]Academic Centre for Biotechnology of Lipid Aggregates, Przybyszewskiego 63-77, 51-148 Wroclaw; [3]University of Zielona Gora, Institute of Biotechnology, Monte Cassino 21b, 65-651 Zielona Gora, Poland

Abstract: Gene therapy is a new approach in the treatment of numerous diseases caused by damage to or lack of certain genes. Its application seems to be very effective, especially in the case of cancer diseases, where genetic drugs can be designed to selectively eliminate damaged cells. Due to the chemical properties of nucleic acids, their introduction into the cell is very difficult. DNA is a highly polar and negatively charged molecule unable to cross the plasma membrane. Moreover, "naked DNA" is exposed to enzymatic degradation by nucleases present within cells and biological fluids. Thus, for efficient and selective DNA delivery into the cells, it is necessary to use a suitable carrier system that can protect it from nucleases and enable it to cross the plasma membrane. Cationic lipid- and synthetic polymer-based carriers are very promising tools for gene delivery into different cell types, but there are still many obstacles to overcome before efficient gene therapy using synthetic vectors will be truly viable. Studies on the nature of interaction between genetic drugs and their lipid carriers, as well as on the mechanism of gene delivery into the cell allowed improvements to be made to currently existing synthetic vectors, and newer and more effective systems to be created. It is crucial that the lipid carriers and genetic drugs have a low degree of toxicity and high degree of specificity for effective gene delivery to occur. That is why the efforts of many laboratories are focused on designing and producing new forms of genetic drugs and derivatives of lipids in order to facilitate the creation of suitable gene delivery systems.

Key words: Cationic lipid, DNA delivery, gene therapy, lipoplex, oligonucleotides.

A list of abbreviations is provided at the end of this chapter.

M.R. Mozafari (ed.), Nanocarrier Technologies: Frontiers of Nanotherapy, 139–174.
© 2006 Springer. Printed in the Netherlands.

1.   INTRODUCTION

Gene therapy is a very intensively investigated strategy for the prevention and treatment of many diseases. Although there are a number of available promising results of *in vitro* tests, many obstacles must be overcome in order to improve selective and effective gene delivery into target cells and tissues. During recent years, cationic lipid-based gene delivery systems have been developed as a tool for efficient gene delivery, but many problems remain, including genetic drug protection against enzymatic degradation, biocompatibility of lipid carriers, specific delivery of the DNA-lipid carrier complex to certain cells or tissues, and satisfactory pharmacokinetics in *in vivo* experiments. Scientists have undertaken many efforts to improve the physiochemical properties of synthetic vectors in order to enhance their *in vivo* efficiency. The main challenge for gene therapy as a strategy in therapeutic treatment is to obtain high transfection efficiency of targeted cells and to reduce any side-effects such as toxicity for untargeted cells and tissues. Another problem that should be resolved is to increase the range of available genetic drugs. In this regard, novel chemical modifications of nucleic acids, yielding promising results in suitability and efficiency assessments have recently been introduced. In this review, we present recent progress and perspectives on the gene delivery field using synthetic non-viral vectors and new genetic drugs. It should be emphasized that understanding the mechanisms of genetic drug delivery using synthetic vectors into target cells and tissues is crucial for constructing a suitable and effective carrier system for gene therapy. The physicochemical natures of the genetic drugs and the lipid compounds of the carrier, as well as the interactions between these molecules are of primary importance. The physicochemical profiles of synthetic vectors determine their fate in biological fluids and after contact with their target cells (uptake and intracellular trafficking), and are important for improving the transfection efficiency of non-viral vector-mediated genetic drug delivery. Within recent years, the efforts of many research groups have focused on new genetic drug development in order to improve their effectiveness and specificity, and to provide the possibility of full control of their action. We present the most promising trends in this field.

2.   GENE THERAPY AND NUCLEIC ACID SYNTHETIC CARRIERS

Gene therapy is a promising strategy that could be used to treat or prevent diseases by changing the expression of a gene, substituting healthy for damaged genes or introducing missing genes into cells. There are two main types of gene therapy: *Somatic*, where the genome is changed but that change is not passed down to future generations, and *Germ*, where those changes will be passed down to future generations. Current research is focused on somatic gene therapy. Molecules such as oligonucleotides can be used either to enhance or inhibit gene expression, or, in combination with other therapeutic approaches, to enhance the effectiveness of non-gene therapy (e.g. with chemotherapeutic agents and radiation treatment for cancer diseases). The potential usefulness of liposomes as a tool for gene transfer has been investigated intensively during recent years [1-5, 7, 8]. The most promising results were obtained in transfection experiments using various cationic liposomes [9-34]. Examples of synthetic cationic lipids which have been tested are shown in Table 9-1.

Liposomes containing various cationic lipids exhibit a diversity of transfection efficiencies, immunogenicities, toxicities and stabilities in the presence of serum. The therapeutic agents in gene therapy can be plasmid DNA, oligonucleotides or siRNA. These molecules are not able to diffuse far from the injection site or to cross barriers such as the endothelium or the blood-brain barrier, as they possess a high negative charge. Moreover, the net negative charge of these particles promotes opsonization and clearance from the circulation by the macrophage system. Naked DNA is very sensitive to degradation by nucleases, which are present in biological fluids. Until recently, gene carriers were mostly of viral type. Despite their high transfection efficiency, viral vectors have serious drawbacks since immunogenicity and insertional mutations cannot be excluded [35]. Non-viral gene carriers have been developed as an alternative [36-42]. Gene transfer vectors should be safe, stable, cost-effective to manufacture in clinically relevant quantities, and capable of efficient and tissue-specific delivery. Non-viral DNA delivery is a multistage process consisting of DNA condensation, cell entry, endosomal escape, nuclear localization and transgene expression.

2.1 Cell entry

The cell surface is covered with so-called glycocalix, among them heparin sulphate proteoglycans. These negatively charged molecules promote electrostatic binding to polycationic particles. It has been reported that extracellular glycosaminoglycans (GAGs) are important factors that affect gene delivery [43-46]. On the one hand, cell membrane-associated glycosaminoglycans mediate the cellular entry of DNA complexes both *in vitro* [43] and *in vivo* [44]; on the other hand, the interaction between the extracellular or secreted glycosaminoglycans and various complexes decreases the level of gene transfer depending on the structure and charge densities of the carriers and GAGs [43, 45, 47]. It has been shown that GAGs may bind to the surface of positively charged complexes, and in some cases GAGs may replace DNA in the complex, resulting in the uptake of GAGs into the cell instead of DNA [46]. Confocal microscopy observations have shown that extracellular GAGs are taken into cells by the cationic carriers, and that they may alter the intracellular behaviour of the complexes [44]. Lipid/DNA aggregates can pass the plasma membrane by fusion, as well. During this process aggregates are subjected to partial or complete lipid/DNA separation and after this, unprotected DNA is degraded in cytosol by nucleases [48]. Incorporation of fusogenic molecules such as peptides or proteins into DNA lipidic carriers can enhance fusion efficiency [48].

Table 9-1. Examples of synthetic cationic lipids used in the preparation of non-viral carries for gene and oligonucleotide delivery.

Abbreviation	Systematic name of the lipid
BGTC	Bis-guanidinium-tren-cholesterol
CTAB	Cetyltrimethylammonium bromide
DC-Chol	3β-[N-(N'N'-Dimethylaminoethane)carbamoyl]cholesterol
DDAB	Dimethyldioctadecylammonium bromide
DMRIE	1,2-Dimirystoyloxypropyl-3-dimethylhydroxyethylammonium bromide
DMTAP	Dimyristoyltrimethyllammonium propane
DODAB	Dioctadecylodimethylammonium bromide
DODAC	Dioleoylodimethylammonium chloride
DOGS	Dioctadecylamidoglycospermine
DORI	1,2-dioleoyloxypropyl-3-dimethylhydroxyethylammonium bromide
DOSPA	2,3-dioleoyl-N-[2-sperminecarboxamidoethyl]-N,N-dimethyl-1-propanaminium
DOSPR	1,3-dioleoyloxy-2-(6-carboxy spermyl)propylamide-4-acetate
DOTAP	1,2-dioleoyloxy-3-(trimethylammonio)propane
DOTIM	1-[2-(oleoyloxy)-ethyl]-2-oleoyl-3-(2-hydroxyethyl)imidazolinum chloride
DOTMA	N-[1-(2,3-dioleoyloxy)propyl]-N,N,N-trimethylammonium chloride
DPPES	Dipalmytoylphosphatidylethanolamidospermine
EDMPC	1,2-Dimyristoyl-sn-glycero-3-ethylphosphocholine chloride
ELMPC	1,2-Dilauroyl-sn-glycero-3-ethylphosphocholine chloride
EOMPC	1,2-Dioleoyl-sn-glycero-3-ethylphosphocholine chloride
GAP-DLRIE	(+/–)-N-(3-aminopropyl)-N,N-dimethyl-2,3-bis(dodecyloxy)-1-propanaminium bromide
SAINT-n	A series of dialkyl pyridinium-alkyl halides
TMAG	N-(alpha-trimethylammonioacetyl)-diodecyl-D-glutamate

2.2   Endosomal escape

Following endocytosis DNA-carrier complexes are entrapped in an intracellular vesicle which delivers the particles to an endosome. In order to avoid DNA degradation by acid hydrolases and nucleases, which would occur in the endosome after fusion with a lysosome, the DNA must escape from the endosome before lysosomal fusion. Nowadays, there are at least 3 known strategies to facilitate endosomal escape:

1- "pH-sensitive liposomes" contain DOPE, which forms a stable bilayer at physiological pH but at pH range of 5-6 forms a hexagonal II structure that destabilizes the membrane and allows the DNA to escape [49-51].

2- "The proton sponge mechanism" involves cationic polymers containing nitrogen atoms that can be protonated (e.g. polyethylenimine – PEI) and which act as a 'proton sponge". They attract the protons in the endosome, leading to diffusion of more protons accompanied by chloride ions into the endosome (Fig. 9-1 shows the scheme of the proton sponge theory). When osmotic pressure becomes high enough to destroy the endosome, the DNA particles are released into the cytosol [6].

3- "Fusogenic peptide", a synthetic peptide containing the 20-aminoterminal amino acid sequence of influenza virus HA; it is stable at pH 7 but the acidic interior of the endosome elicits fusion with the endosomal membrane, and DNA particles are released into the cytosol.

2.3   Nuclear localization

It appears that mitosis is the critical step for a DNA complex to become located in the nucleus. It is hypothesized that a DNA complex is dragged along with chromosomes that are passing by during mitosis, and incidentally included into the nucleus. In resting cells, the nuclear envelope is the barrier for access to the nucleus. It is unlikely that molecules will pass through a nuclear pore because of size limitations: the pore diameter is 26 nm while the radius of most plasmids is at least 90 nm. In order to improve nuclear localization, nuclear localization signal peptides (NLS) can be chemically incorporated into DNA carriers. NLS contains a short sequence of basic amino acids and forms a stable complex with cytosolic factors karyopherin alpha and beta. The NLS complex docks at nuclear pore proteins and is then translocated to the nucleus.

2.4   Transgene expression

This stage requires suitable promoter, initiator and cleavage sequences. Although early steps including cationic lipid/DNA complex internalization through an endocytosis-like mechanism have been relatively well characterized [52-54], subsequent steps including intracellular complex dissociation and DNA entry into the nucleus are poorly understood. Cornelis *et al.* [32] demonstrated that cationic lipid/plasmid DNA complex dissociation is not a limiting step, and that it can occur both in the cytoplasm and in the nucleus of HeLa cells. However, plasmid DNA released into cytosol could not enter the nucleus and did not result in significant transfection [34]. Cationic lipid/plasmid DNA complex trafficking via the endosomal compartment could be involved in DNA entry into the nucleus but the mechanism of this process is still unknown [34].

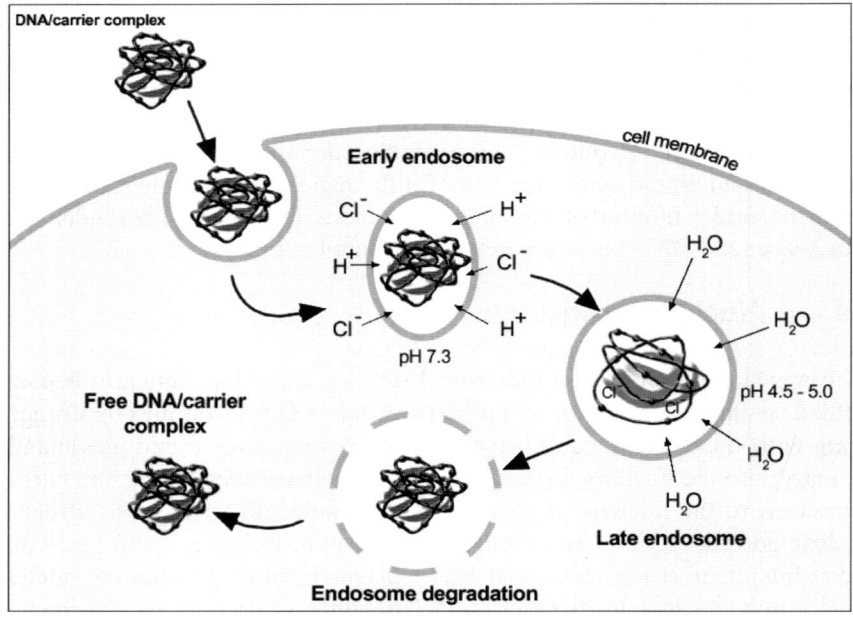

Figure 9-1. The scheme of the proton sponge theory. The DNA/carrier complex enters the cytoplasm by the endocytosis pathway. H^+ and Cl^- ions enter the endosome; Cl^- ions penetrate the complex and the DNA carrier's effective size increases. As the pH of the endosome decreases, H_2O molecules enter the endosome by osmosis. Finally, the DNA is released from the endosome.

It has been shown that a synthetic vector must have a strong positive charge to hold high quantities of DNA [6]. It was recently reported that lipoplexes formulated from highly fluorinated lipids are very promising synthetic DNA vectors. Available experimental data has shown that lipoplexes composed of highly fluorinated analogues of DOGS and highly fluorinated helper lipids, analogues of dioleoylohosphatidylethanolamine (DOPE), exhibit a higher *in vitro* and *in vivo* transfection potential than conventional lipoplexes [38, 55-57]. The unique hydrophilic and hydrophobic character of these lipids prevents DNA from interactions with lipophilic and hydrophilic biocompounds and from degradation [38].

3. THE TOOLS FOR ANTISENSE GENE THERAPY: SYNTHETIC GENE TRANSFER VECTORS

It is commonly known that carrier systems, mainly lipid based ones, are required for the effective introduction of genetic drugs into cells. Synthetic vectors have proved able to improve transfection efficiency and therapeutic factor (such as antisense oligonucleotide) stability in both *in vitro* and *in vivo* experiments. Recently, numbers of new synthetic lipid carriers have been developed and tested and a number of these have been shown to enhance the efficiency of gene delivery and transfection. Their improved properties enabled an increase in genetic drug stability and cell and tissue specificity, and a decrease in side effects. Designing tools for gene therapy in order to optimise gene delivery does not only mean working with lipid compounds; it also involves performing chemical modifications of the genetic drugs.

The antisense oligonucleotides (asDNA, asODN) are single-stranded DNA molecules containing a complementary sequence directed against the target gene. They are designed to hybridise with a specific sequence of mRNA. This hybridization induces RNase activity, or sterically inhibits mRNA translation. In order to deliver asODN to a cell, and to protect them from enzymatic degradation, various liposomal formulations have been designed, but the commonly used ODN carriers consist of pre-formed cationic vesicles with or without DOPE mixed with ODN [58, 59].

3.1 Liposomes

It is known that positively charged lipid aggregates incorporate ODN (and other nucleic acids) very efficiently and are able to deliver it to the cells *in vitro* [58-60]. Disadvantages of these complexes are the large size and excess positive charge, which result in unfavourable pharmacokinetics in *in vivo* tests. It has been shown that cationic lipoplexes are not able to deliver ODN to tissues other than MPS (mononuclear phagocytic system) and lung

[8]. In many cases, cationic lipoplexes are not able to maintain their surface charge and colloidal stability in the presence of blood, due to interaction with negatively charged blood components. Anionic liposomes can be an alternative to cationic DNA carriers. Unfortunately, they are not free of flaws either. Anionic liposomes cannot form complexes with nucleic acids or passively encapsulate naked DNA with high efficiency because DNA is a negatively charged molecule. In order to overcome this obstacle, a newer formulation was designed, known as LPD II (liposome-polycation-DNA complex), where DNA is complexed with cationic polymers (such as protamine [61]) followed by mixing with fusogenic anionic liposomes. Since LPD II particles more closely resemble natural viral particles, they should transfect targeted cells with a higher degree of efficiency. Moreover, these types of gene carrier are less toxic and even less immunogenic than cationic ones. Cationic lipid delivery systems possess a net positive charge that induces unspecific interactions with blood components leading to opsonization, destabilization and rapid uptake by phagocytes [41, 62-64]. A novel type of liposomal vector for gene therapy, namely AVPs (Artificial Virus Particles), was prepared using DLPE, DOPS and cholesterol (the major constituents of the HIV envelope [65]). The DNA is condensed using low molecular weight, branched PEI. It was shown that this type of DNA carrier is serum-resistant, non-toxic and small enough to be endocytosed [41]. Additionally, in order to improve AVP selectivity for tumour endothelial and melanoma cells, liposomes were equipped with RGD peptide (cyclic peptide). RGD motif acts as a ligand for $\alpha_v\beta_3$-integrin, which is a surface marker in tumour endothelial and melanoma cells [66, 67].

A method for ODN entrapment in neutral liposomes is *passive encapsulation* within their aqueous space [68, 69]. Although the advantage of this type of carrier is the small size and the lack of a positive charge that would induce opsonization by plasma protein and MPS uptake, the incorporation efficiency, especially in small liposomes (<200 nm), is not satisfactory [68, 70]. It has been reported that the addition of 2-5 mol% of PEG-DSPE into neutral liposome passively encapsulated ODNs results in an increased degree of incorporation efficiency [58]. In order to increase the half-life of these neutral vesicles in circulation, polyethylene glycol can be grafted onto their surface leading to a reduction in MPS uptake [71-75].

Another example of ODN carrier is Coated Cationic Liposomes (CCL), which can be obtained using a procedure similar to that for plasmid DNA encapsulation [76, 77]. This method is based on the fact that cationic lipids such as DOTAP could be used for the extraction of plasmid DNA from an aqueous into an organic phase. This process was shown to be mediated by electrostatic interactions between the positively charged lipid and the negatively charged DNA. This method can be applied to negatively charged

ODN. Complexes obtained in the organic phase are then coated with neutral lipids through a reverse evaporation step [58]. In this method, it was shown that increasing the amount of DOTAP increases the ODN extracted from the aqueous into the organic phase. The aqueous phase in which ODN is dissolved plays an important role in the extraction efficiency. It was reported that at a DOTAP/ODN charge ratio of 0.88, more than 80% of the ODN is extracted from the ddH$_2$O aqueous phase, and approximately 70% when ODN is dissolved in 10% sucrose. On the other hand, the extraction is completely inhibited in the presence of 25 mM Hepes, 140 mM NaCl (pH 7.4) [58]. A charge-shielding of Na$^+$ and Cl$^-$ ions prevents efficient interaction between DOTAP and ODN. Another important parameter in this system is the amount of coating lipids; this affects the ability for CCL extrusion but has no effect on the ODN incorporation efficiency [58].

Cationic liposomes could also be used as a carrier for immunostimulatory CpG oligonucleotide delivery [78]. These ODNs are promising immune adjuvants, anti-allergens and immunoprotective agents. It was reported that encapsulation in SSCL (sterically stabilized cationic liposomes) protects ODN from degradation, increases their uptake by cells of the immune system and significantly enhances their immunostimulatory activity *in vitro* and *in vivo* when compared with free ODNs [55-57, 78]. The advantages of this type of DNA lipid carriers for cancer chemotherapy have been demonstrated [79-83]. Although viral vectors display numerous drawbacks for *in vivo* application, efforts were undertaken to reduce their side-effects by combining the properties of viral- and non-viral DNA carriers. It was demonstrated that the conjugation of cationic liposomes – composed of TMAG (Table 9-1), DLPC (dilauroyl phosphatidyl-choline) and DOPE (dioleoyl phosphatidyethanolamine) – with adenoviral (Ad) vectors containing herpes simplex virus thymidine kinase results in the reduction of viral antigenicity, while antitumour activity is maintained. This approach for gene therapy seems to be very promising for human cancer treatment, especially for malignant gliomas [84]. Research was performed on combination non-viral DNA vectors and viral components and on many other tools capable of enhancing targeted delivery and transfection efficiency [85-87]. In order to improve targeted delivery of antisense drugs new procedures for suitable carrier systems are being developed.

3.1.1 Immunoliposomes

Immunolipoplexes based on a lipoplex backbone (consisting of cationic lipids complexed with oligonucleotides) and equipped with targeting elements (antibodies), were shown to be a good non-viral vector for specific oligonucleotide transfer into human lymphoma cells, which are known to be a very difficult object for transfection [88-90].

An alternative tool for specific therapeutic gene delivery *in vivo* could be PEGylated Immunoliposomes (PIL) [91]. In this type of formulation, DNA is encapsulated in the interior of neutral or anionic liposomes [92-94] that are equipped with specific antibodies (conjugated e.g. via polyethylene glycol chain) as a targeting ligand.

3.2   Cationic emulsions

The other approach to obtain a new nucleotide drug carrier is "submicron cationic emulsion" [95, 96]. It was proven that positively charged O-W (oil in water) emulsion is safe and tolerated in both intravenous and ocular administration [97]. Oligonucleotides up to 50 mer in length can be efficiently associated with the oil droplets of this emulsion by ion-pair formation at the O-W interface. Despite this, ODN molecules were shown to be protected from nuclease degradation in cell culture conditions [96, 98]. Recently, novel non-viral delivery systems consisting of nanoparticles formulated from biodegradable polymers such as poly(lactic acid) (PLA) and poly(lactide-co-glycolide) (PLGA) are under intensive investigation. PLGA nanoparticles for DNA delivery are mainly formulated by an emulsion-solvent evaporation technique using PVA as a stabilizer generating negatively charged particles and heterogeneous size distribution. Nevertheless, cationically modified PLGA nanoparticles have been developed [99]. It has been demonstrated that these types of vector possess defined size and shape, and as they are able to bind DNA very efficiently, thus could serve as a potential tool for effective gene transfer.

4.   OPTIMIZATION OF LIPID CARRIER PARAMETERS IN ORDER TO IMPROVE TRANSFECTION EFFICIENCY

The precise mechanisms of gene delivery into cells are still unknown. For years, many efforts were made in order to attain a step-by-step understanding of this process. Our knowledge in this field is still growing. Designing and optimizing new forms of lipid-based gene delivery vectors is possible thanks to the new information that is being discovered and processed.

For DNA transfer into cells, several important conditions must be fulfilled. DNA has to be protected against enzymatic degradation, the DNA/carrier complex must be internalized into the cell and the functionality of the DNA has to be preserved for the expression of the gene in the cell. In order to improve the transfection efficiency of lipid DNA carriers, several modifications can be introduced. Optimizing the cationic lipid/DNA charge ratio and precondensating plasmid DNA with poly-L-lysine or adding polyethylene glycol (PEG) in the transfection medium can enhance the transfection efficacy of non-viral DNA carriers [35]. One of the most important parameters that affect their activity is lipid composition [100]. It is well known that the targeted delivery of oligonucleotide drugs shows varying efficiency depending on the cell type and the nature of the carrier. The advantages of cholesterol-based cationic liposomes as a tool for brain cancer gene therapy [101, 102] and their superiority over other cationic lipids in experiments *in vivo* [103] were demonstrated. Cholesterol is a known stabilizer of the liposomal membrane and is able to modulate its fluidity [104], but the introduction of some structural modifications into the cholesterol molecule can change its chemical properties. It has been reported that the ramified quaternary ammonium polar group of TEAPC-Chol (3-β [N-(N'N'N'-triethyl aminopropane) carbamoyl] cholesterol iodide) will not be affected by the pH or charged components in the transfection media. It forms stable liposomes when mixed with the helper lipid DOPE [35]. These cationic vesicles bind negatively charged plasmid DNA as a result of electrostatic interactions. In the presence of serum, a greater amount of TEAPC-Chol/DOPE is required for complete complexation of DNA because of the competition between the negatively charged serum components and DNA in the binding of TEAPC-Chol liposomes. The cationic lipid/DNA charge ratio is a key point for transfection. When the liposome bilayer structure is conserved, the charge ratio lip^+/PO_2^- should be 2, because only half of the lipid molecules are in the outer side of the liposomes interacting with the DNA. When the liposome structure is destroyed, all the cationic lipid molecules would be associated with the DNA, and the charge ratio would be 1 [35]. Based on this, a molar ratio in the range of 1 to 2 probably leads to the formation of a heterogeneous distribution of complexes [35, 105, 106]. The correlation between transfection levels in the cells and the different structure of the complex resulting from the variable molar charge ratio has been reported on [15, 35]. The value of the optimum charge ratio for an efficient transfection depends on the nature of the lipid vector, the carried DNA, and the cell line. Cytosol entry is the first important step for liposome-mediated transfection. It has been demonstrated that a low concentration of PEG could induce liposome aggregation, increasing the association between the cell membrane and the lipid/DNA complex. This improved association leads to better penetration of the complex into the cytoplasm [35, 107]. Another limiting step for liposome-mediated DNA delivery is nuclear transport of the DNA following cytoplasmic penetration. Recent observation revealed that poly-L-lysine can mimic a nuclear localisation signal (NLS) and increase the ability of DNA to enter the nucleus [108]. It is also proposed that DNA condensed by poly-L-lysine is protected against enzymatic degradation in the cytosol [35]. Although, polylysine condenses DNA very efficiently, the resulting structure is hydrophilic. It is obvious that in order to cross the cell membrane, highly hydrophilic DNA molecules must be "hidden" inside some hydrophobic structure. Recently, a new, very interesting approach for DNA complexation with cationic compounds has been presented. The combination of two condensing agents (DOTAP) and PLL (polylysine), differing in their affinity towards water, when mixed with plasmid, resulted in aggregates which are resistant to enzymatic digestion and possess well-defined size distribution [109]. Such complexes, in which polylysine ensures very efficient DNA condensation and DOTAP provides hydrophobicity may be further modified (outer lipid layer formation) in order to obtain a fully functional carrier for genetic material [109].

It was demonstrated that colloidal carriers were able to improve the cellular uptake of charged molecules such as oligonucleotides. As mentioned above, a very promising strategy for DNA delivery is a system of synthetic vectors consisting of cationic lipids [110-112]. Unfortunately, a problem that can be observed in many cases is an aggregation of these charged particles, especially when prepared at a high concentration or in the presence of tissue culture media [25, 113]. At a critical lipid/DNA ratio, aggregation, precipitation or even fusion can occur [114]. Due to the biological barriers that must be crossed by lipidic carriers, such as the endothelium or the blood-brain barrier, it is essential to unify particle size below 1 μm. A common method to homogenise the formed aggregates and to achieve process-controlled size homogeneity on a laboratory scale is the use of sonication [110]. This can cause DNA fragmentation which is dependent on the lipid composition of the liposomal vectors. The ultrasonic cavitation destroys DNA molecules either directly due to the physical forces focused on the molecules or indirectly by highly reactive sonochemicals such as hydrogen peroxide [115]. This leads to single and double strand cleavage, rupture of hydrogen-bonds, base destruction and possibly cross-link formation [116]. On the other hand, certain liposome formulations are able to protect bound plasmid from damage by sonication [113]. One example of such a formulation is DEAE-stabilized polyhexylcyanoacrylate (PHCA)-nanoparticles. In this case, after 10 min sonication, 80% of the adsorbed oligonucleotides remained intact [110]. Moreover, ODN damage caused by ultrasonication was found to be independent of any chemical modification of the oligonucleotide backbone. Although short sonication times (30-60 s) seem to be acceptable for the homogenization of ODN-liposome complexes, both unmodified phosphodiester oligonucleotides and phosphorothioates were degraded in a time-dependent manner to an extent of 60% within 10 minutes [110]. Another approach to improve ODN stability during the carrier preparation procedure, in the presence of serum supplemented culture media and in circulation, is to use ODNs in complex with protamine; the ODN/protamine complex is named proticles [117]. Protamine spontaneously forms complexes with ODNs, which aggregate to negatively charged particles with a diameter of 46-219 nm. The size of these complexes depends on protamine/ODN ratios. The diameters of 51 and 2.5:1 proticles ranged between 46-138 nm, the 1:1 particle diameter was between 81-219 nm and diameter of the 0.5:1 aggregates between 67-100 nm [117]. After complexation of proticles with AH-Chol-liposome, particle size increased, resulting in positively charged complexes of a diameter of 350-400 nm [118]. Moreover, protamine increases the binding efficiency and the stability of asODNs bound to cationic AH-Chol liposomes [119]. Additionally, such carriers exhibit a greater degree of protection of ODNs against ultrasonic cavitation and a higher stability in cell culture medium than AH-Chol-liposomes alone [117].

5. pH-SENSITIVE LIPOSOMES FOR GENETIC DRUG DELIVERY AND THE ROLE OF DOPE AS A HELPER LIPID

In the search for new solutions in the field of gene delivery optimization, several attempts were made using natural phenomena. One example is the viral infection process. It is well known that a number of viral infections involve the fusion of the viral membrane with the endosome membrane as soon as the pH of this compartment starts to decrease [120, 121], which is why those natural systems are so efficient at delivering genetic material directly to the cytosol even if they reach the cytoplasm via the endocytotic pathway.

Conventional liposomes mainly enter the cell via endocytosis. However, the major obstacle for liposome-mediated drug delivery is the slow rate of drug release and lack of fusogenic activity of conventional vesicles following internalization into the endosomes. In answer to this issue, pH-sensitive liposomes were designed for oligonucleotide delivery. This type of carrier is designed to be stable at physiological pH and destabilized in an acidic environment, i.e. following cellular internalization, promoting the leakage of their content into the cytosol [122-124]. Since it has been recognized that *in vivo* membrane fusion may involve a hexagonal organization of phospholipids, vesicles would undergo a lamellar to hexagonal H_{II} phase in the acidic environment of the endosome, and fuse with the endosomal membrane, enabling delivery of their content into the cytoplasm, thus avoiding enzymatic degradation [120]. The bilayer of pH-sensitive liposomes contains phosphatidylethanolamine (PE) [125], which forms a non-bilayer H_{II} phase in water but is able to organize in a lamellar phase when mixed together with other lipids. In liposomal structure, the transition of PE from a lamellar to a hexagonal phase is affected by temperature, salinity and pH [126]. At neutral pH, headgroup ionization provides sufficient electrostatic repulsion to stabilize the phospholipids in a lamellar phase. However, the design of PE liposomes that would be stable in biological conditions requires the use of secondary components containing titratable acidic groups which provide electrostatic repulsion preventing the formation of the hexagonal phase at physiological pH [120]. In commonly used PE liposomes, the stabilizers are fatty acids like oleic acid [127], cholesterol hemisuccinate (CHEMS) [128], or palmitoylhomocystein [125]. DOPE possesses a strong propensity to form a non-bilayer structure due to its cone-shaped geometry. CHEMS is weakly acidic amphiphile which confers stability of the bilayer phase at neutral pH [19, 129-131]. At acidic pH, CHEMS becomes partially protonated and loses its negative charge; therefore, its ability to stabilize the bilayer structure based on electrostatic repulsion results in destabilisation and/or liposome fusion [122]. Although it has been shown that DOPE-based liposomes enhance the cytosolic delivery of many therapeutic agents [132-134], their *in vivo* application is limited by their relatively poor stability in the presence of serum [135-137]. The addition of polyethyleneglycol (PEG)-derivatives of PE to DOPE-liposomes enhanced their stability in the presence of serum, but significantly reduced pH-sensitivity [138]. Another helper lipid that can be used to enhance transfection efficiency is DOPC (dioleoylphosphatidylcholine) and in some cases is even more effective then DOPE [139].

Recently, the formulation of novel, pH-sensitive liposome containing oleyl alcohol (OAlc) in combination with egg phosphatidylcholine (PC) as the membrane-destabilizing component was tested [122]. It is suggested that OAlc, an unsaturated fatty alcohol, is capable of forming a hydrogen bond through its hydroxyl group to an oxygen atom on the phosphate group on the PC molecule, resulting in the formation of a complex with a geometry similar to that of DOPE (Fig. 9-2). This results in a lowering of the energy barrier for the lipid transition from a lamellar to a hexagonal II phase, which is required for membrane destabilisation [122]. OAlc-containing liposomes were shown to be very stable and pH-sensitive in the presence of serum. Moreover, the degree of pH-sensitivity can be modulated by altering the OAlc content in the liposome formulation [122]. Additionally, the incorporation of Tween-80 results in increasing liposome stability but reducing pH-dependent aggregation [122]. The optimization of the OAlc/Tween-80 ratio might improve the procedures of pH-sensitive liposome preparation. These types of lipid carriers may be applied in delivering antisense oligodeoxyribonucleotides or plasmid DNA in gene therapy. The mechanism by which pH-sensitive liposomes induce fusion after phase transition from a lamellar to a hexagonal phase is not completely clear. The possible explanations are: 1) mixing of the aqueous content and mixing of phospholipidic components (complete fusion); or 2) mixing of bilayer components without leakage or mixing of the aqueous content (incomplete fusion); or 3) mixing of bilayer components without mixing of the aqueous content but with leakage of the aqueous content (lysis) [120]. The best mechanism for oligonucleotide delivery from lysosomes into the cytoplasm is complete fusion; however, this process is complicated and depends on the composition of the pH-sensitive liposomes [120].

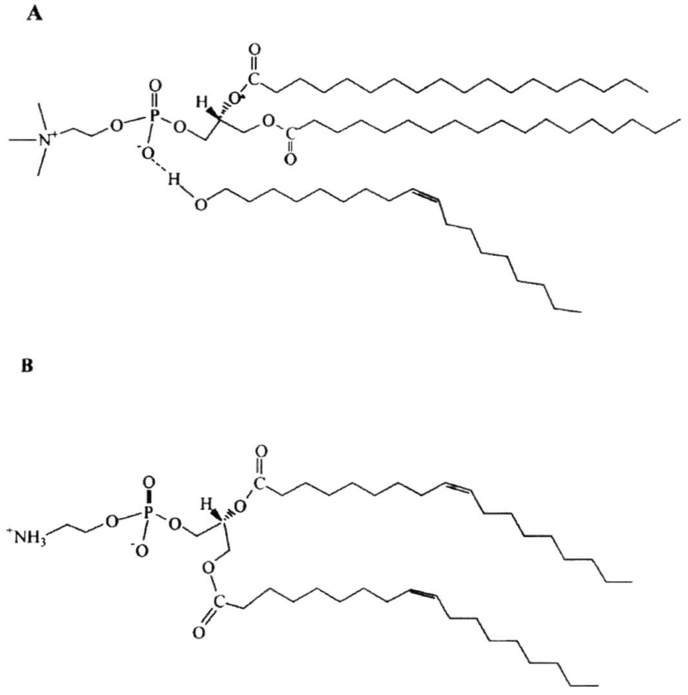

Figure 9-2. A comparison of the proposed PC/OAlc complex (A) and DOPE (B). In both structures, the hydrophilic zones of the molecule occupy a relatively small volume compared to the lipophilic region. This should promote the formation of non-bilayer structures and liposome destabilization [122].

The pH-sensitive immunoliposomes show a multi-step mechanism of cytoplasmic delivery consisting of: 1) specific binding to the cell surface; 2) internalization into acidic endosomes; 3) fusion of the liposome with the endosomal membrane; and/or 4) endosomal rupture; and 5) release of the liposomal content into the cytosol [120].

6.   ANTISENSE OLIGONUCLEOTIDE MODIFICATIONS

Effectiveness of gene therapy requires suitable genetic drugs that possess a high specificity of action that would be convenient to control and monitor. Of the range of genetic drugs that could be used in gene therapy, antisense nucleotides seem to be among the most promising, and they can be employed as effective gene-specific regulators. Essentially, all oligos are large polar molecules which are not able to cross the cell membrane to an appreciated extent due to their charged polyanionic phoshodiester backbone.

Under normal cell culture conditions, antisense oligos enter cultured cells by endocytosis and are degraded or remain sequestered in lysosomes, or they are exocytosed back into the culture medium. It is thus necessary to achieve a significant level of delivery into the cytosol/nuclear compartment simultaneously with protection from the enzymatic activity of the nucleases present in cells and biological fluids. So far phosphorothioate oligos (S-DNA) have dominated the antisense field due to their high transfection efficiency in serum-free systems and improved resistance to nuclease activity. However, S-DNAs have poor sequence specificity, probably because of their RNase H-based mechanism of action [140]. Moreover, S-DNAs bind many proteins in the extracellular milieu, on the cell surface and within cells, resulting in multiple non-antisense effects [140]. S-DNAs are degraded by nucleases in serum and in cells over a period of hours to several days, so their effects are relatively transient. Additionally, degradation of S-DNA within cells releases thioated nucleotides which are likely to be toxic [140]. Another difficulty is the prediction of the effective target for S-DNA. Usually, 6-12 S-DNA oligos complementary to a given mRNA must be prepared and tested in order to find one which is effective in cells. For several years, as an alternative to phosphorothioate derivatives of oligo-nucleotides, morpholino antisense nucleotides have been tested [141-150]. Recently, GENE TOOLS began commercial production of morpholino oligos, which effectively overcome the multiple limitations inherent in S-DNA. The comparison of morpholino and phosphorothioate structures and a short characterisation of their action are presented in Fig. 9-3 and Table 9-2.

Figure 9-3. The chemical structures of morpholinos (left) and phosphorothioates (right) [140].

Table 9-2. A comparison of morpholino and phosphorothioate oligonucleotides [140].

MORPHOLINOS	PHOSPHOROTHIOATES
Predictable targeting	Unpredictable targeting
Reliable activity in the cells	Unreliable activity in cells
Minimal non-antisense activity	Multiple non-antisense activities
Very high sequence specificity	Poor sequence specificity
Complete resistance to nucleases	Degraded by nucleases
Fast, simple and reliable delivery	Complicated delivery

7.   A NEW APPROACH TO GENE THERAPY (siRNA and PNAs)

7.1   siRNA

One of the newest trends in genetic drug research is the application of the recently discovered phenomenon of RNAi (RNA interference). This is a powerful experimental tool for reducing the expression of specific genes. That is why it can be exploited to create new, gene-specific therapeutics.

RNAi is a process in which the introduction of double-stranded RNA (dsRNA) into cells causes degradation of the complementary mRNA (www. ambion.com/techlib-/append/RNAi_mechanism.html). In the cell, long dsRNA (typically >200 nucleotides) is cleaved into short 20-25 nucleotide small interfering RNA (siRNA) by a ribonuclease called Dicer (an RNase III-like enzyme). The siRNAs subsequently assemble with protein components into a RNA-induced silencing complex (RISC). This complex contains endoribonuclease. The activated RISC then binds to complementary sequence on mRNA molecule by base-pairing interaction. The bound mRNA is cleaved near the middle of the siRNA binding region (Fig. 9-4). Sequence specific degradation of mRNA results in gene silencing [151-153].

Long double-stranded RNA (dsRNA) can be used to silence the expression of target genes in different organisms (e.g. worms, fruit flies, plants, mice); however, the introduction of dsRNA longer than 30 nucleotides into mammalian cells initiates a potent antiviral response, exemplified by non-specific inhibition of protein synthesis and RNA degradation. This problem can be overcome by introducing siRNA instead of dsRNA into the cell.

In addition to their role in gene silencing, siRNA have been determined to play diverse biological functions *in vivo*, including antiviral defence, transpozon silencing, gene regulation, centromeric silencing and genomic rearrangements (www.ambion.com/techlib/append/RNAi_mechanism.html).

7.2   PNAs (peptide nucleic acids)

Antisense technology is very promising for therapeutic applications, but the efficiency of antisense oligonucleotides remains limited by their rapid degradation by intracellular nucleases, insufficient target affinity, non-specific side effects and their inefficient uptake due to the low permeability of the cell membrane [154-157]. In order to overcome these obstacles, novel oligonucleotide chemistries have been developed, and several new generations of antisense nucleotides have been proposed in the literature [158, 159], particularly peptide nucleic acids (PNAs), which seem to be a very promising tool for antisense gene therapy in both eukaryotic and prokaryotic cells [157, 160, 161].

In PNAs, the phosphodiester backbone of DNA or RNA is replaced by a hybrid molecule corresponding to N-(2-aminoethyl)glycine monomers linked by amide bonds [162, 163].

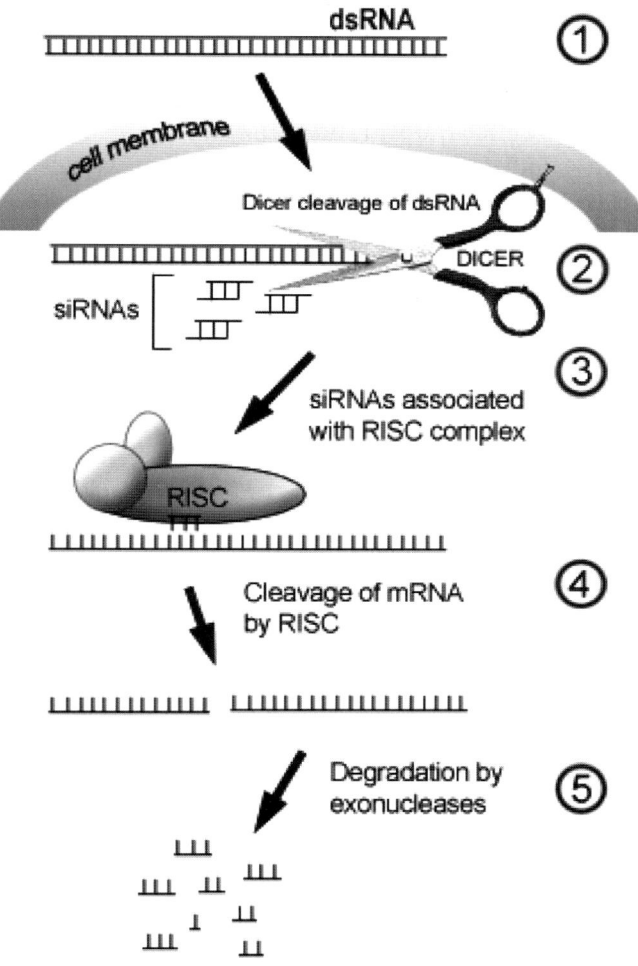

Figure 9-4. The mechanism of RNA interference (RNAi).

PNAs allow specific gene targeting, are highly stable, are resistant to nuclease and protease, and bind RNA or DNA targets in a sequence-specific manner with high affinity [157]. However, only a few reports actually associate a strong biological response with the use of PNAs *in vivo*, which is in part related to their poor propensity to cross the cell membrane and/or to their inappropriate cellular localization [164]. In order to improve delivery of PNAs into cells, several chemical modifications based on covalently bound cell-penetrating peptides have been successfully applied [164-168]. The linkage of nuclear localization or polylysine sequences to PNAs have also been found to improve both their cellular internalization and nuclear translocation [157, 169-171]. It was recently shown that the cellular uptake of PNAs depends on cell type, and to achieve efficient uptake, a carrier system is necessary [43, 168]. The solubility of PNAs and their tendency to self-aggregate are important factors to be considered prior to their selection for biological use [172-174]. PNA solubility is often sequence-dependent. This generally gives rise to several limitations in PNA probe design; this situation can be improved by incorporation of lysine residue or polyethylene glycol linkers [170, 172-175].

Recently, a novel oligonucleotide that mimics a dimeric oligomer consisting of a phosphonate analogue of PNA (pPNA) and a PNA-like monomer based on a *trans*-4-hydroxyl-L-proline (Hyp-NA) [176, 177] was discovered. The introduction of negative charges into the PNA backbone provides excellent solubility without hindering the other properties. The HypNA-pPNA chimera interacts with the target sequence in the same manner as classical PNAs, independently of ionic strength [157, 176-179]. Recently, a new technology that combines a noncovalent peptide-based delivery system with a new generation of PNAs or HypNAs was described [157]. In order to demonstrate the potential of this technology, the authors chose as a target a protein that is essential for cell cycle progression, cyclin B1 [157]. They designed a peptide carrier called Pep-2 [179]. Pep-2-mediated the delivery of an antisense HypNA-pPNA chimera directed specifically against cyclin B1, leading to a rapid and strong downregulation of cyclin B1 and efficiently blocking the cell cycle progression of several cell lines, including that derived from a breast cancer [157]. They also demonstrated that, by contrast to antisense oligonucleotides, PNAs did not trigger endogenous RNaseH activity, since only molecules whose structure is closest to that of DNA promote the activation of this enzyme [159, 160]. The mechanism through which PNAs causes downregulation of gene expression seems to be associated with steric hindrance, as PNA binding to mRNA is known to disrupt ribosome assembly, and block target protein synthesis [154, 157, 158, 160].

8. PROPOSED MECHANISM OF CATIONIC LIPIDS-DNA COMPLEXATION

Understanding the mechanisms of cationic lipids/oligonucleotides complexation is fundamental to optimizing parameters that are crucial for effective cell transfection, such as a high quantity of complexed oligonucleotides, their ability to dissociate from the complex after cell entry, and their stability. The recognition of processes of DNA/cationic lipid complexation and the behaviour of such formulates in the biological environment/*in vivo* would facilitate the designation and formation of new and effective gene delivery systems with improved transfection properties.

The interaction of cationic liposomes with polyanions in aqueous solution results in the spontaneous formation of aggregates of various sizes ranging from 100-500 nm to micrometer size [180-182]. So far, little is known about the relationship between the structure, properties and morphology of such complexes and their transfection efficiency. In order to understand the nature of interactions which mediate the formation of these complexes numerous experiments have been performed [180]. It was found that relative concentration of two stable populations of structures in the 100-500 nm and 5 μm range, depended on the salt concentration [180]. Electrostatic interactions play a key role in determining the physico-chemical properties of DNA-lipid aggregates. For charge stabilized colloidal suspensions, the aggregation process is governed within the Derjaguin-Landau-Verwey-Overbeek (DLVO) model [180, 183, 184] by the repulsive energy barrier in the interaction potential between two approaching particles. If the height of this barrier is less than the thermal energy k_BT, every collision will result in the particle sticking together and very rapid aggregation, limited only by the rate of diffusion-induced collision between the clusters (DLCA aggregation). When the energy barrier is of an order or higher than k_BT, many collisions must occur before two particles can stick together; thus the aggregation rate is much lower (RLCA aggregation) [180]. By contrast, in cationic liposome suspensions, aggregation and in some cases fusion can be induced by the reduction of electrostatic repulsion between particles caused by the binding of counterions to the charged lipid surface. In this case, at least two different cluster populations can occur; their relative concentrations depend on the ionic strength [180]. The DLVO theory [185] could explain how the parameters such as size, surface charge and zeta-potential affect the final properties of aggregates [113]. It is obvious that the net charge density of the liposome surface is not constant, due to the presence of ions in the solution binding to their surface and modifying their charge [180]. According to DLVO theory for charged colloidal stability, the interactions between two approaching particles are governed by long-range van der Waals and long-range electrostatic repulsion forces. In agreement with the potential profile, particle collision can lead to rapid or slow coagulation. Within the DLVO theory, the stability of liposomes depends on the salt concentration, the counterion valency, the surface potential Φ_o, the liposome diameter 2R and the Hamaker constant A_H (Fig. 9-5 shows a typical profile for the interaction potential between two approaching liposomes).

The stability of liposomes is ensured by the presence of a potential barrier higher than the thermal energy k_BT. The addition of salt causes a reduction in its height and initiates the aggregation process [180]. Among other factors that should be taken into account to characterise the vesicle aggregation phenomenon are the fluctuation of surface geometry and hydration forces [186]. These parameters have not been regarded in DLVO theory. The aggregation process promoted by simple monovalent salt in cationic liposomes suspensions differs from that caused by polyvalent salt (in particular polynucleotides), and in this case, complex formation cannot be simply described by DLVO theory alone [180].

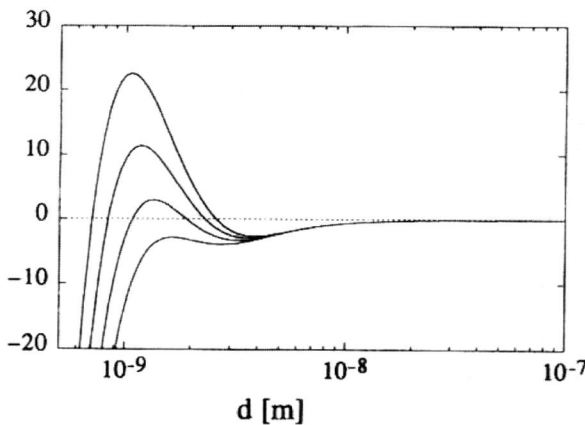

Figure 9-5. A typical profile of interaction between two approaching liposomes for different values of the surface potential Φ (from top to bottom) (a) $\Phi = 120$mV; (b) $\Phi = 100$mV; (c) $\Phi = 90$mV. The particle size diameter is 2R = 80 nm and the Hamaker constant $A_H = 0.195 \times 10^{-21}$ J (Reprinted from Ref. [180] with permission from Elsevier).

The release of counter-ion entropy upon complexation is the driving-force that mediates DNA-cationic lipid complexation [187, 188]. Due to this energy being very high (of the order of 1 kT/A) it is expected that DNA always forms some type of complex with cationic lipids [187]. The difference in the free energy of DNA-lipid aggregates determines their relative stability. This energy is composed of head-to-head interaction energy, tail-to-tail interaction energy and interfacial energy. Both the head-to-head and tail-to-tail interaction energies are the function of surface density, alone [187, 189, 190]. In a lipid-DNA complex, this parameter is a function of DNA concentration because of the electrostatic screening effect [191]. The surface density of the inner and outer layers in a cylindrical DNA-lipid complex will be different as well. The third contribution of the free energy of aggregated lipids is provided by the deformation energy of the hydrophobic tails (associated with perturbation from an optimal length [187, 190]. Suitable calculations facilitate the prediction of which structure should be the most stable.

9.   POSSIBLE STRUCTURES OF DNA-CATIONIC LIPOSMES COMPLEXES

It is well known that various liposomal DNA carrier systems show different transfection efficiencies depending on the lipid composition or cell lines [192]. It is hypothesised that DNA complex geometry will influence the interaction between complexes and the cell, causing different transfection effectiveness [90, 187, 193]. Experimental data suggests three possible geometries of DNA/cationic liposome complexes [64, 187, 194, 195] (Fig. 9-6).

The simplest structure, which is a "bead on a string", arises when positively charged multilamellar liposomes adhere on the negatively charged DNA [64]. Lamellar complexes represent spherical macro-aggregates composed of flat lipid bilayers with DNA molecules packed between them in a "sandwich"-like structure [187]. These macro-aggregates can be joined by a DNA strand [187]. Cylindrical formulations occur when cationic lipids coat the DNA strand with a cylindrical-shaped bilayer [187]. Although the existence of all three formations has been reported [64, 187, 191, 194, 195], some authors claim that the only stable geometry of DNA-lipid aggregates is multi-lamellar one, regardless of lipid properties [187] with the exception of lipids which form hexagonal II phases [187, 189]. In this case DNA is coated by a single monolayer rather than a bilayer. Such a structure, named a "honeycomb" is more stable than a cylindrical one [196].

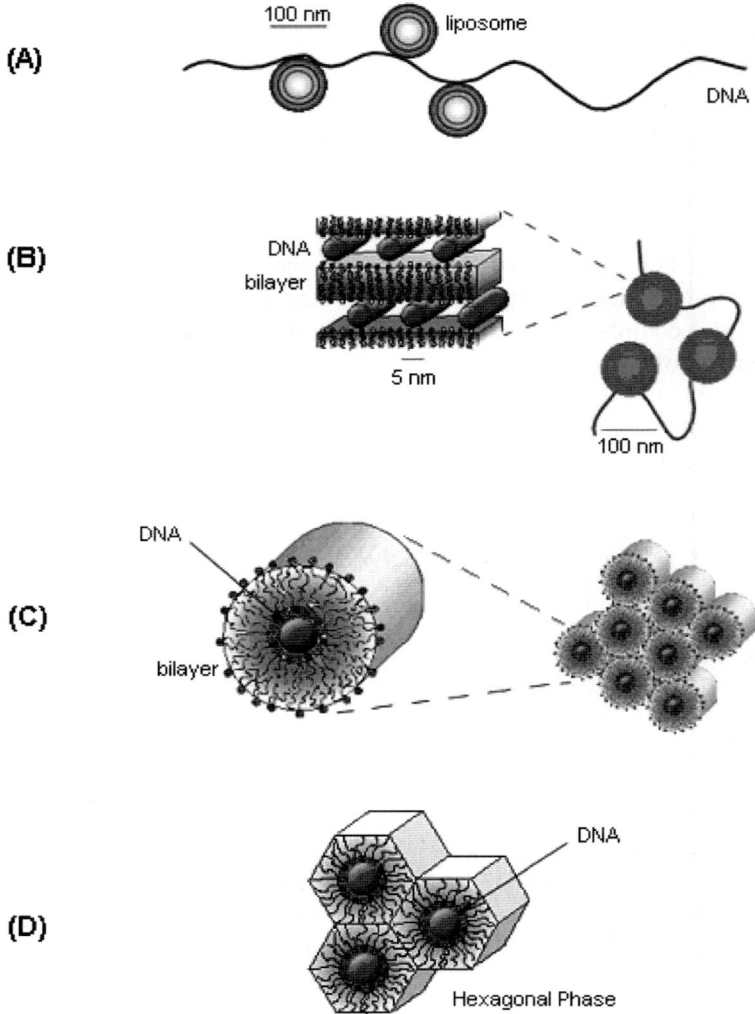

Figure 9-6. The structures of DNA complexes with cationic liposomes (modified from [187]): A. "bead on a string" complexes, B. lamellar complexes, C. cylindrical complexes, D. honeycomb structures.

10. APPLICATION OF FLUORESCENCE CORRELATION SPECTROSCOPY TECHNIQUE FOR DNA CONDENSATION STUDIES

Understanding the nature of interactions between DNA molecules and cationic compounds is crucial for predicting structures of such complexes and is very important for designing effective, synthetic carriers for genetic drugs. Aggregate formation can be considered as a two-stage process: neutralization of DNA phosphate group negative charges with positively charged compounds, causing DNA condensation and subsequent aggregate rearangement determining complex topology [197, 198]. DNA condensation can be caused by ions, proteins, cationic lipids and polymers but the mechanisms of these processes are still unknown. Difficulties in controlling the DNA condensation process results in heterogenous and unstable aggregates [197]. Studying of such multi-component systems is the most credible when two approaches are employed: whole sample analysis and single molecule studies [197]. The FCS technique provides unique possibility of measuring a statistical distribution of particles that differ in their hydrodynamic size, by accumulating a large number of single molecule measurements. FCS is based on measurements of the residence time of labeled DNA molecules in a small volume, where excitation time is focused [197]. This technique can be used to measure intermolecular interactions such as ligand-receptor binding, protein and polymer aggregation and the diffusion of labeled molecules on the lipid bilayer plane or within cell [197]. It has been demonstrated that the differences in hydrodynamic sizes of relaxed and condensed DNA are high enough to follow DNA condensation by using FCS technique [197, 199]. Moreover by using FCS one is able to detect the differences between various condensing agents (e.g. HTAB (hexadecyl-trimethylammonium bromide) [197], spermine [199] or DOTAP [200]), hence this technique can be very useful in designing new synthetic carriers for genetic drug delivery.

11. CONCLUDING REMARKS

Cationic lipid- and synthetic polymer-based carriers are very promising tools for genetic drugs delivery into different cell types, but there are still many obstacles to overcome before efficient gene therapy using synthetic vectors will be really possible. Studies on the nature of interaction between genetic drugs and their lipid carriers, as well as on the mechanism of gene delivery into the cell allow improvements of existing synthetic vectors and the development of new systems.

Acknowledgement
Authors thank to Mr Derek Handley for language correction and Prof. Marek Langner from Wroclaw University of Technology for critical reading of the manuscript.

List of abbreviations

AH-Chol: (cholest-5-en-3b-yl-6-aminohexyl ether), **asODN**: antisense nucleotides, **AVPs**: artificial virus particles, **CCL**: cationic coated liposome, **CHEMS**: cholesterol hemisuccinate, **DLPC**: dilauroyl phosphatidylcholine, **DLPE**: dilauroyl phosphatidylethanolamine, **DOPE**: dioleoyl phosphatidylethanolamine, **DOPC**: dioleoyl phosphatidylcholine, **DOPS**: dioleoyl phosphatidyl serine, **dsRNA**: double-stranded RNA, **DVLO**: Derjaguin-Landau-Verwey-Overbeek theory, **GAGs**: glycosaminoglycans, **HTAB**: hexadecyl trimethyl ammonium bromide, **Hyp-NA**: *trans*- 4- hydroxyl- L- proline, **LPD**: liposome-polycation-DNA complex, **MPS**: mononuclear phagocytic system, **NLS**: nuclear localization signal, **OAlc**: oleyl alcohol, **ODN**: oligonucleotides, **PC**: phosphatidylcholine, **PEI**: polyethylenimine, **PEG-DSPE**: 1,2- distearoyl- sn-glycero- 3-phospho ethanolamine- N- [poly (ethyleneglycol) 2000], **PHCA**: DEAE- stabilized polyhexyl- cyanoacrylate, **PIL**: pegylated immunoliposome, **PLA**: poly (lactic acid), **PLGA**: poly (lactide-co-glycolide), **PLL**: polylysine, **PNAs**: peptide nucleic acids, **pPNA**: phosphonate analogue of PNA, **RISC**: RNA-induced silencing complex, **RNAi**: RNA interference, **S-DNA**: phosphorothioate oligos, **siRNA**: small interfering RNA, **TEAPC-Chol**: 3-β [N- (N'N'N'- triethyl aminopropane) carbamoyl] cholesterol iodide.

12. REFERENCES

1. Itani T., Ariga H., Yamaguchi N., Tadakuma T., Yasuda T. (1987) A simple and efficient liposome method for transfection of DNA into mammalian cells grown in suspension. Gene 56: 267-276.
2. Kaneda Y., Uchida T., Kim J., Ishiura M., Okada Y. (1987) The improved efficient method for introducing macromolecules into cells using HVJ (Sendai virus) liposomes with gangliosides. Exp. Cell Res. 173: 56-59.
3. Mannito R.T., Goulg-Fogerite S. (1988) Liposome mediated gene transfer. Biotechniques 6: 682-690.
4. Haga N., Yaki K. (1989) An improved method for entrappment of plasmids in liposomes. J. Clin. Biochem. Nutr. 7: 175-183.
5. Kato K., Nakanishi M., Kanede Y., Uchida T., Okafa Y. (1991) Expression of hepatitis B virus surface antigen in adult rat liver: co-introduction of DNA and nuclear protein by a simplified method. J. Biol. Chem 266: 3361-3364.
6. Pang S.W., Park H.Y., Jang Y.S., Kim W.S., Kim J.H. (2002) Effects of charge density and particle size of poly(styrene/(dimethylamino)ethyl methacrylate) nanoparticle for gene delivery in 293 cells. Colloids and Surface B: Biointerfaces 26: 213-222.
7. Ito K., Asano T., Vaughan E.D. Jr, Poppas D.P., Hayakawa M., Felsen D. (2004) Liposome-mediated gene therapy in kidney. Hum. Cell 17: 17-28.
8. Litzinger David C., Brown Jeffrey M., Wala I., Kaufman Stephen A., Van Gwyneth Y., Farrell Catherine L., Collins David (1996) Fate of cationic liposomes and their complex with oligonucleotide *in vivo*. Biochim. Biophys. Acta 1281: 139-149.
9. Zhao Dan-Dan, Watarai Shinobu, Lee Jin-Tae, Kouchi, Ohmori Hitishi, Yasuda Tatsuji (1997) Gene transfection by cationic liposomes: comparision of the transfection efficiency of liposomes prepared from various positively charged lipids. Acta Med. Okayama 51: 149-154.

10. Felgner P.L., Gadek T.R., Holm M., Roman R., Chan H.S., Wenz M., Northrop J.P., Ringold G.M., Danielsen H. (1987) Lipofection: A highly efficient, lipid-mediated DNA transfection procedure. Proc. Natl. Acad. Sci. USA 84: 7413-7417.

11. Brigham K.L., Meyrick B., Berry L.C., King G. Jr (1989) Expression of prokaryotic gene in cultured lung endothelial cells after lipofection with a plasmid vector. Am. J. Respir. Cell. Mol. Biol. 1: 95-100.

12. Koshizaka T., Harashi Y., Yaki K. (1989) Novel liposomes for efficient transfection of B-galactoside gene into COS-1 cells. J. Clin. Biol. Chem. Nutr. 7: 185-192.

13. Felgner P.L. (1990) Particulate systems and polymers for *in vitro* and *in vivo* delivery of polynucleotides. Adv. Drug Delivery Rev. 5: 167-187.

14. Muller S.R., Sullivan Pd., Clegg D.O., Feinstein S.C. (1990) Efficient transfection and expression of heterologous gene in PCI 2 cells. DNA Cell. Biol. 9: 221-229.

15. Gao X., Huang L. (1991) A novel cationic liposome reagent for efficient transfection of mammalian cells. Biochem. Biophys. Res. Commun. 179: 280-285.

16. Yagi K., Noda H., Kurono M., Ohishi N. (1993) Efficient gene transfer with less cytotoxicity by means of cationic multilamellar liposomes. Biochem. Biophys. Res. Commun. 196: 1042-1048.

17. Wolff J.A., Malone R.W., Williams P., Chong W., Ascadi G., Jani A., Felgner P.L. (1990) Direct gene transfer into mouse muscle *in vivo*. Science 247: 1465-1468.

18. Rose J.k., Buonocore L., Whitt M.A. (1991) A new cationic liposome reagent mediating nearly quantitative transfection of animal cells. Biotechniques 10: 520-525.

19. Litzinger D.C., Huang L. (1992) Phosphatidylethanolamine liposomes: drug delivery, gene transfer and immunodiagnostic applications. Biochim. Biophys. Acta 1113: 201-227.

20. Stribling R., Beunett E., Liggitt D., Gaensler K., Debs R. (1992) Aerosol gene delivery *in vivo*. Proc. Natl. Acad. Sci USA 89: 11277-11281.

21. Zhu N., Liggitt D., Liu Y., Debs R. (1993) Systemic gene expression after intravenous DNA delivery into adult mice. Science 261: 209-211.

22. Ohilip R., Liggitt D., Philip M., Dazin P., Debs R. (1993) *In vivo* gene delivery: efficient transfection of T lymphocytes in adult mice. J. Biol. Chem. 268: 16087-16090.

23. Plautz G.E., Wu B., Gao X., Huang L., Nabel G.T. (1993) Immunotherapy of malignancy by *in vivo* gene transfer into tumours. Proc. Natl. Acad. Sci USA 90: 4645-4649.

24. Trivedi R.A., Dickson D. (1995) Liposome-mediated gene transfer into normal and dystrophin-deficient mouse myoblasts. J. Neurochem. 64: 2230-2238.

25. Hofland H., Huang L. (1995) Inhibition of human ovarian carcinoma cell proliferation by liposome-plasmid DNA complex. Biochem. Biophys. Res Commun. 207: 492-496.

26. Kojima H., Ohishi N., Takamori M., Yagi K. (1995) Cationic multilamellar liposome-mediated gene transfer into primary myoblast. Biochim. Biophys. Res. Commun. 207: 8-12.

27. Meyer K.B., Thompsom M.M., Levy M.Y., Barron L.G., Szoka F.C. Jr (1995) Intratracheal gene delivery to the mouse airway: characterization of plasmid DNA expression and pharmacokinetics. Gene Ther. 2: 450-460.

28. Gaucheron J., Boulanger C., Santaella C., Sbirrazzidi N., Boussif O., Vierling P. (2001) *In vitro* cationic lipid-mediated gene delivery with fluorinated glycerophospho-ethanolamine helper lipids. Bioconjug. Chem. 12: 949-963.

29. Rejman J., Wagenaar A., Engberts J.B., Hoekstra D. (2004) Characterization and transfection properties of lipoplexes stabilized with novel exchangeable polyethylene glycol- lipid conjugates. Biochim. Biophys. Acta 28: 41-52.

30. Yingyongnarangkul B.E., Howarth M., Ekkiott T., Bradley M. (2004) Solid phase synthesis of 89-polyamine-based cationic lipids for DNA delivery to mammalian cells. Chemistry 23: 463-473.

31. Van Zanten J., Doornbos-Van der Meer B., Audouy S., Kok R.J. de Ley L. (2004) A nonviral carrier for targeted gene delivery to tumor cells. Cancer Gene Ther 11: 156-164.

32. Cornelis S., Vandenbranden M., Ruysschaert J-M., Elouahabi A., (2002) Role of intracellular cationic lipid-DNA complex dissociation in transfection mediated by cationic lipids. DNA and Cell Biology 21: 91-97.

33. Karmali P.P., Kumar V.V., Chaudhuri A. (2004) Design, synthesis and *in vitro* gene delivery efficacies of novel mono-, di- and trilysinated cationic lipids, a structure-activity investigation. J. Med. Chem. 47: 2123-2132.

34. Montier T., Delepine P., Benvegnu T., Ferrieres V., Miraman M.L., Dagorn S., Guillaume C., Plusquellec D., Ferec C. (2004) Efficient gene transfer into human epithelial cell lines using glycosylated cationic carriers and neutral glycosylated co-lipids. Blood Cell Mol. Dis. 32: 271-282.

35. Lesage D., Cao A., Briane D., Lievre N., Coudert R., Raphael M., Salzmann J.L., Taillandier E. (2002) Evaluation and optimization of DNA delivery into gliosarcoma 9L cells by a cholesterol-based cationic liposome. Biochim. Biophys. Acta 1564: 393-402.

36. Schatzlein A.G. (2001) Non-viral vectors in cancer gene therapy:principles and progress. Anticancer Drugs 12: 275-304.

37. Gregoriadis G. (1999) DNA vaccines: a role for liposomes. Curr. Opin. Mol. Ther. 1: 39-42.

38. Gaucheron J., Santaella C., Vierling P. (2002) Transfection with fluorinated lipoplexes based on fluorinated analogues of DOTMA, DMRIE and DPPES. Biochim. Biophys. Acta 1564: 349-358.

39. Felgner P.L., Barenholz Y., Behr J.P., Cheng S.H., Cullis P., Huang L., Jessee J.A., Seymour L., Szoka F., Thierry A.R., Wagner E., Wu G. (1997) Nomenclature for synthetic gene delivery systems. Hum. Gene Ther. 8: 511-512.

40. Verderone G., Van Craynest N., Boussif O., Santaella C., Bischoff R., Kolbe H.V.J., Vierling P. (2000) Lipopolycationic telomers for gene transfer: synthesis and evaluation of their *in vitro* transfection efficiency. J. Med. Chem. 43: 1367-1379.

41. Fahr A., Muller K., Nahde T.H., Muller R., Brusselbach S. (2002) A new collidal lipidic system for gene therapy. J. Liposome research 12: 37-44.

42. Erni C., Suard C., Freitas S., Dreher D., Merkle H.P., Walter E. (2002) Evaluation of cationic solid lipid microparticles as synthetic carriers for the targeted delivery of macromolecules to phagocytic antigen-presenting cells. Biomaterials 23: 4667-4676.

43. Mislick K.A., Baldechwieler D. (1996) Evidence fort he role of proteoglycans in cation-mediated gene transfer. Proc. Natl. Acad.Sci. USA 93: 12349-12354.

44. Mounkes L.C., Zhong W., Cipres-Palaci G., Heath T.D., Debs R.J. (1998) Proteoglycans mediate cationic liposome-DNA complex-based gene delivery *in vitro* and *in vivo*. J. Biol. Chem. 273: 26164-26170.

45. Ruponen M., Yla-Herttuala S., Urtti A. (1999) Interaction of polymeric and liposomal gene delivery systems with extracellular glycosaminoglycans: physicochemical and transfection studies. Biochim. Biophys. Acta 1415: 331-341.

46. Ruponen M., Ronkko S., Honkakoski P., Tammi M., Urtti A. (2001) Extracellular glycosaminoglycans modify cellular trafficking of lipoplexes and polyplexes. J. Biol. Chem 276: 33875-33880.

47. Belting M., Petersson P. (1999) Intacellular accumulation of secreted proteoglycans inhibits cationic lipid-madiated gene transfer. Co-transfer of glycosaminoglycans to the nucleus J. Biol. Chem. 274: 19375-19382.

48. Langner M. (2000) The intracellular fate of non-viral DNA carriers. Cell. Mol. Biol. Lett. 5: 295-313.

49. Straubinger R.M., Duzgunes N., Papahadjopoulos D. (1984) pH-sensitive liposomes mediate cytoplasmic delivery of encapsulated macromolecules. FEBS Lett. 179: 148-154.

50. Tachinaba R., Futaki S., Harashima H., Kiwada H. (2003) pH-sensitive Liposomes in nuclear targeting of macromolecules. Methods In Enzymology 372: 349-361.

51. Oudrhiri N., Vigneron J.-P., Peuchmaur M., Lecrerc T., Lehn J.-M., Lehn P. (1997) Gene transfer by guanidinium-cholesterol cationic lipids into airway epithelial cells *in vitro* and *in vivo*. Proc. Natl. Acad. Sci USA 94: 1651-1656.

52. Zuhorn I.S., Kalicharans R., Hoekstra D. (2002) Lipoplex-mediated transfection of mammalian cells occurs through the cholesterol-dependent clathrin-mediated pathway of endocytosis. J. Biol. Chem. 277: 18021-18028.

53. El Ouahabi A., Thiry M., Pector V., Fuks R., Ruysschaert J.M., Vandenbranden M. (1997) The role of endosome destabilizing activity in the gene transfer process mediated by cationic lipids. FEBS Lett. 414: 187-192.

54. El Ouahabi A., Thiry M., Schiffmann S., Fuks R., Nguyen-tran H., Ruysschaert J.M., Vandenbranden M. (1999) Intracellular visualization of BrdU-labeled plasmid DNA/cationic liposome complexes. J. Histochem. Cytochem. 47: 1159-1166.

55. Gaucheron J., Santaella C., Vierling P. (2001) Highly fluorinated lipospermines for gene transfer: synthesis and evaluation of their *iv vitro* efficiency. Bioconjug. Chem. 12: 114-128.

56. Gaucheron J., Santaella C., Vierling P.(2001) Imprioved *in vitro* gene transfer mediated by fluorinated lipolexes in the presence of a bile salt surfactant. J. Gene Med. 3: 338-344.

57. Boussif O., Gaucheron J., Boulanger C., Santaella C., Kolbe H.V.J., Vierling P. (2001) Enhanced *in vitro* and *in vivo* cationic lipid-mediated gene delivery with a fluorinated glycerophospho-ethanolamine helper lipid. J. Gene Med. 3: 109-114.

58. Stuart D.D., Allen T.M. (2000) A new liposomal formulation for antisense oligodeoxynucleotides with small size, high incorporation efficiency and good stability. Biochim. Biophys. Acta 1463: 219-229.

59. Bennett C.F., Chiang M., Chan H., Shoemaker J.E., Mirabelli C.K. (1992) Cationic lipids enhance cellular uptake and activity of phosphorothioate antisense oligonucleotides. Mol. Pharmacol. 41: 1023-1033.

60. Zelphati O., Szoka F.C. (1996) Intracellular distribution and mechanism of delivery of oligonucleotides mediated by cationic lipids Pharm. Res. 13: 1367-1372.

61. Sun X., Zhang Z. (2004) Optimizing the novel formulation of liposome-polycation-DNA complex (LDP) by central composite design. Arch. Pharm. Res. 27: 797-805.

62. Vile R.G., Sunassee K., Diaz R.M., (1998) Strategies for achieving multiple layers of selectivity in gene therapy. Molecular Medicine Today 4: 84-92.

63. Fisher D., Bieber T., Li Y., Elsasser H.-P., Kissel T.A. (1999) A novel non-viral vector for DNA delivery based on low molecular weight, branched polyethylenimine:effect of molecular weight on transfection efficiency and cytotoxicity. Pharmaceutical research 16: 1273-1279.

64. Sternberg B., Sorgi F.L., Huang L. (1994) New structures in complex formation between DNA and cationic liposomes visualized by freeze-fracture electron microscopy. FEBS Lett. 356: 361-366.

65. Chander R., Schreier H. (1992) Artificial envelopes containing recombinant human immunodeficiency virus (HIV) gp160. Life Sci. 50: 481-489.

66. Nahde T., Muller K., Fahr A., Muller R., Brusselbach S. (2001) Combined transductional and transcriptional targeting of melanoma cells by artificial virus-like particles. J. Gen. Med. 3: 353-361.

67. Pasqualini R., Koivunen E. (1997) Ruoslathi Symbol.av integrins as receptors for tumor targeting by circulating ligands. Nature Biotech. 15: 542-546.
68. Leonetti J., Machy P., Degols G., Lebleu L., Leserman (1990) Antibody-targeted liposomes containing oligodeoxyribonucleotides complementary to viral RNA selectively inhibit viral replication Proc. Natl. Acad. Sci USA 87: 2448-2451.
69. Thierry A.R., Dritschilo A. (1992) Intracellular availability of unmodified phosphorothioated and liposomally encapsulated oligodeoxynucleotides for antisense activity. Nucleic Acids Res. 20: 5691-5698.
70. Ropert C., Malvy C., Couvreur P. (1993) Inhibition of the Friend retrovirus by antisense oligonucleotides encapsulated in liposomes: mechanism of action. Pharm. Res. 10: 1427-1433.
71. Klibanov A.L., Maruyama K., Torchilin V.P., Huang L. (1990) Amphipathic, poly-ethyleneglycols effectively prolong circulation time of liposomes. FEBS Lett. 268: 235-237.
72. Papahadjopoulos D., Allen T.M., Gabizon A., Mayhew E., Matthay K., Huang S.K., Lee M.C., Woodle D.D., Lasic C., Redemann F.J.M. (1991) Sterically stabilized liposomes: improvements in pharmacokinetics and antitumor therapeutic efficacy. Proc. Natl. Acad. Sci. USA 88: 11460-11464.
73. Ahmad I., Longenecker J., Samuel J., Allen T.M. (1993) Antibody-targeted delivery of doxorubicin entrapped in sterically stabilized liposomes can eradicate lung cancer in mice. Cancer Res 53: 1484-1488.
74. Mayhew E., Allen T.M., Newmwn M.S., Woodle M.C., Vaage J., Uster P.S. (1995) Pharmacokinetics and antitumor activity of vincristine encapsulated in sterically stabilized liposomes. Int. J. Cancer 62: 199-204.
75. Lopez de Menezes D.E., Pilarski L.M., Allen T.M., (1998) *In vitro* and *in vivo* targeting of immunoliposomal doxorubicin to human B-cell lymphoma. Cancer Res. 58: 3320-3330.
76. Reimer D.L., Zhang Y.P., Kong S., Wheeler J.J., Graham R.W., Bally M.B. (1995) Formation of novel hydrophobic complexes between cationic lipids and plasmid DNA. Biochemistry 34: 12877-12883.
77. Wong F.M.P., Reimer D.L., Bally M.B. (1996) Cationic lipid binding to DNA: characterization of complex formation. Biochemistry 35: 5756-5763.
78. Gursel I., Gursel M., Ishii K.J., Klinman D.M. (2001) Sterically stabilized cationic liposomes improve the uptake and immunostimulatory activity of CpG oligonucleotides. Journal Immunol. 167: 3324-3328.
79. Gregoriadis G. (1998) Genetic vaccines: strategies for optimalization. Pharmacol Res. 15: 661.
80. Whitmore M., Huans L.S., (1999) LPD lipopolylex initiates a potent cytokine response and inhibit tumor growth. Gene Ther 6: 1867.
81. Meyer O., Kirpotin D., Hong K., Sternberg B., Park J.W., Woodle M.C., Papahadjopoulos D. (1998) Cationic liposomes coated with polyethylene glycol as carriers for oligonucleotides. J. Biol. Chem. 273: 15621.
82. Lasic D.D., Papahadjopoulos D. (1995) Liposomes revisited. Science 267: 1275.
83. Hong K., Zheng W., Baker A., Papahadjopoulos D. (1997) Stabilization of cationic liposome-plasmid DNA complexes by poliamines and poly(ethylene glycol)-phospholipid conjugates for efficient *in vivo* gene delivery. FEBS Lett. 400: 233.
84. Mizuno M., Ryuke Y., Yoshida J. (2002) Cationic liposomes conjugation to recombinant adenoviral vectors containing herpes simplex virus thymidine kinase gene followed by ganciclovir treatment reduces viral antigenicity and maintains antitumor activity in mouse experimental glioma model. Cancer Gene Ther. 9: 825-829.

85. Hwang S.H., Hayashi K., Takayama K., Maitani Y. (2002) Liver-targeted gene transfer into a human hepatoblastoma cell line and *in vivo* by sterylglucoside-containing cationic liposomes. Gene Ther. 8: 1276-1280.
86. Heikkila A., Hiltunen M.O., Turunen M.P., Keski-Nisula L., Turunen A.M., Rasanen H., Rissanen T.T., Kosma V.M., Kosma V.M., Manninen H., Heinonen S., Yla-Herttuala S. (2001) Angiographically guided utero-placental gene transfer in rabbits with adenoviruses, plasmid/liposomes and plasmid/plyethyleneimine complexes. Gene Ther. 8: 784-788.
87. Murray K.D., Etheridge C.J., Shah S.I., Matthews D.A., Russell W., Gurling H.M., Miller A.D. (2001) Enhanced cationic liposome-mediated transfection using the DNA-binding peptide mu(mu) from the adenovirus core. Gene Ther. 8: 453-460.
88. Guillem V.M., Tormo M., Moret I., Benet I., Garcia-Conde J., Crespo A., Alino S.F. (2002) Targeted oligonucleotide delivery in human lymphoma cell lines using a polyethyleneimine based immunolipolexes. J. Control Release 83: 133-146.
89. Guillem V.M., Tormo M., Revert F., Benet I., Garcia-Conde J., Crespo A., Alino S.F. (2002) Polyethyleneimine- based immunopolyplex for targeted gene transfer in human lymphoma cell lines. J. Gene Med. 4: 170-182.
90. Smisterowa J., Wagenaar A., Stuart M.C.A., Polushkin E., ten Brinke G., Hulst R., Engberts J.B.F.N., Hoekstra D. (2001) Molecular shape of the cationic lipid controls the structure of cationic lipid/dioleylphosphatidylethanolamine- DNA complexes and the efficiency of gene delivery. J. Biol. Chem. 276: 47615-47622.
91. Pardridge W.M. (2003) Gene targeting *in vivo* with pegylated immunoliposomes, Methods in Enzymology 373: 507-528.
92. Shi N., Pardridge W. M. (2000) Noninvasive gene targeting to the brain, Proc. Natl Acad. Sci. USA 97: 7567-7572.
93. Shi N., Boado R.J., Pardridge W.M., (2001) Receptor-mediated gene targeting to tissue *in vivo* following intravenous administration of pegylated immunoliposomes, Pharm Res. 18: 1091-1095.
94. Shi N., Zhang C., Zhu C., Boado R.J., Pardridge W.M. (2001) Brain-specific expression of an exogenous gene after *in vivo* administration Proc. Natl. Acad. Sci. USA 98: 12754-12759.
95. Tae Woo Kim, Hesson Chung, Ick Chan Kwon, Ha Chin Sung, Seo Young Jeong. (2001) Optimization of lipid composition in cationic emulsion as *in vitro* and *in vivo* transfection agents. Pharm. Res. 18:54-60.
96. Teixeira H., Dubernet C., Rosilio V., Laigle A., Deverre J.R., Scherman D., Benita S., Couvreur P. (2001) Factors influencing the oligonucleotides release from O-W submicron cationic emulsion. J. Control. Release 70: 243-255.
97. Klang S., Frucht-Pery J., Hoffman A., Benita S. (1994) Physicochemical characterization and acute toxicity evaluation of a positively-charged submicron emulsion vehicle. J. Pharm. Pharmacol. 46: 986-993.
98. Teixeira H., Dubernet C., Chacun H., Rabinovich L., Boulet V., Deverre J.R., Benita S., Couvreur P. (2003) Cationic emulsion improves the delivery of oligonucleotides to leukemic P388/ADR cells in ascite. J Control release. 20: 473-482.
99. Ravi Kumar M.N., Bakowsky U., Lehr C.M. (2004) Preparation and characterization of cationic PLGA nanospheres as DNA carriers. Biomaterials 25: 1771-1777.
100. Langner M. (2000) Effect of liposomal composition on its ability to carry drug. Pol. J. Pharmacol. 52: 3-14.
101. Natsume A., Mizuno M., Ryuke Y., Yoshida J. (1999) Antitumor effect and cellular immunity activation by murine interferon-beta gene transfer against intracerebral glioma in mouse. Gene Ther. 6: 1626-1633.

102. Yagi K., Ohishi N., Hamada A., Shamoto M., Ohbayashi M., Ishida N., Nagata A., Kanazawa S., Nishikimi M. (1999) Basic study on gene therapy of human malignant glioma by use of the cationic multilamellar liposome-entrapped human interferon beta gene. Hum. Gene Ther. 10: 1975-1982.

103. Egilmez N.K., Iwanuma Y., Bankert R.B., (1996) Evaluation and optimization of different cationic liposome formulations for *in vivo* gene transfer. Biochim. Biophys. Res. Commun. 221: 169-173.

104. Pang K.Y., Miller K.W. (1978) Cholesterol modulates the effects of membrane perturbes in phospholipids vesicles and biomembranes. Biochim. Biophys. Acta 511: 1-9.

105. Cao A., Briane D., Courdert R., Vassy J., Lievre N., Olsman E., Tamboise E., Salzmann J.L., Rigaut J.-P., Tailliandier E. (2000) Delivery and pathway in MCFt cells of DNA vectorized by cationic liposomes derived from cholesterol. Acid. Drug Dev. 10: 369-380.

106. Zabner J., Fasbender A.J., Moninger T., Poellinger K.A., Welsh M.J. (1995) Cellular and molecular barriers to gene transfer by a cationic lipid. J. Biol. Chem. 270: 18997-19007.

107. Yang Q., Guo Y., Li L., Hui S.W. (1997) Effect of lipid headgroup and packing stress on poly(ethylene glycol) induced phospholipids vesicle aggregation and fusion. Biophys J. 73: 277-282.

108. Zauner W., Brunner S., Buschle M., Ogris M., Wagner E. (1999) Differential behaviour of lipid based and polycation based gene transfer systems in transfecting primary human fibroblast: a potential role of polylysine in nuclear transport. Biochim. Biophys. Acta 1428: 57-67.

109. Chaszczewska-Markowska M., Ugorski M., Langner M. (2004) Plazmid condensation induced by cationic compounds: hydrophilic polylysine and amphiphilic cationic lipid. Cell. Mol. Biol. Lett. 9: 3-13.

110. Zobel H.-P., Werner D., Gilbert M., Noe C.R. Stieneker F., Kreuter J., Zimmer A. (1999) Effect of ultrasonication on the stability of oligonucleotides adsorbed on nanoparticles and liposomes. J. Microencapsulation 16: 501-509.

111. Bertling W.M., Gareis M., Paspaleeva V., Zimmer A., Kreuter J., Nurnberg E., Harrer P. (1991) Use of liposomes, viral capsids and nanoparticles as DNA carriers. J. Biotechnol. Appl. Biochem. 13: 390-405.

112. Cherng J.Y., Van De Wetering P., Talsma H., Crommelin D.J., Hennink W.E. (1996) Effect of size and serum proteins on transfection efficiency of poly((2-dimethylamino)ethyl methacrylate)-plasmid nanoparticles. Pharm. Res. 13: 1038-1042.

113. Wasan E.K., Reimer D.L., Bally M.B. (1996) Plasmid DANN is protected against ultrasonic cavitation-induced damage when complexed to cationic liposomes. J. Pharm. Sci 85: 427-433.

114. Solodin I., Brown C.S., Bruno M.S., Chow C.Y., Jang E.H., Debs R.J., Heath T.D. (1995) A novel series of amphiphilic imidazolinium compounds for *in vitro* gene delivery. Biochemistry 34: 13537-13544.

115. Kondo T., Arai S., Kuwabara M., Yoshi G., Kano E. (1985) Damage in DNA irradiated with 1.2 MHZ ultrasound and its effect on template activity of DNA for RNA synthesis. Radiation Research 104: 284-292.

116. Munshi C.B., Graeff R., Lee H.C. (2002) Evidence for a causal role of CD38 expression in granulocytic differentiation of human HL-60 cells. J. Biol. Chem. 277: 49453-49458.

117. Junghans M., Kreuter J., Zimmer A. (2000) Antisense delivery using protamine-oligonucleotide particle. Nucleic Acids Res. 28: 45.

118. Zimmer A., Atmaca-Abdel Aziz S., Gilbert M., Werner D., Noe C.R. Synthesis of cholesterol modified cationic lipids for liposomal drug delivery of antisense oligonucleotides. Eur. J. Pharm. Biopharm. 47: 175-178.

119. Sorgi F.L., Bhattacharaya S., Huang L. (1997) Protamine sulfate enhances lipid-mediated gene transfer. Gene Ther. 4: 961-968.

120. Fattal E., Delattre J., Dubernet C., Couvreur P. (1999) Liposomes for delivery of nucleotides and oligonucleotides. S.T.P. Pharma Science 9: 383-390.

121. White J., Kielian M., Helenius A. (1983) Membrane fusion proteins of enveloped animal viruses. Quart. Rev. Biophys. 16: 151-195.

122. Sudimack Jennifer J., Guo Wenjin, Tjarks Werner, Lee Robert J. (2002) A novel pH-sensitive liposome formulation containing oleyl alcohol. Biochim. Biophys. Acta 1564: 31-37.

123. Torchilin V.P., Zhou F., Huang L. (1993) pH-sensitive liposomes. J. Liposome Res. 3: 201-255.

124. Duzgunes N., Straubinger R.M., Baldwin P.A., Papahadjopoulos D. (1991) in: J. Wilschut, D. Hoekstra (Eds.), Membrane fusion, Marcel Dekker, New York, pp. 713-730.

125. Connor J., Yatvin M.B., Huang L. (1984) pH sensitive liposomes: acid-induced liposome fusion. Proc. Natl. Acad. Sci. USA 81: 1715-1718.

126. Harlos K., Eibl H. (1981) Hexagonal phase in phospholipids with saturated chains: phosphatidylethanolamine and phosphatidic acid. Biochemistry 20: 2888-2892.

127. Duzgunes N., Sraubinged R.M., Baldwin P.A., Friend D.S., Papahadjopoulos D. (1985) Proton induced fusion of oleic acid-phosphatidylethanolaminr liposomes. Biochemistry 24: 3091-3098.

128. Bentz J., Ellens H., Laim M.Z., Szoka jr F.C. (1985) On the correlation between H_{II} phase and the contact induced destabilization of phosphatidylethanolamine-containing membranes. Proc. Natl. Acad. Sci. USA 82: 5742-5745.

129. Liu D., Huang L. (1990) pH-sensitive, plasma-stable liposomes with relatively prolonged resistance in circulation. Biochim. Biophys. Acta 1022: 348-354.

130. Ellens H., Bentz J., Szoka F.C. (1984) pH-induced destabilization of phosphatidylethanolamine-containing liposomes: role of bilayer contact Biochemistry 23: 1532-1538.

131. Cullis P.R., De Kruijff B. (1979) Lipid polymorphism and the functional roles of lipids in biological membranes. Biochim. Biophys. Acta 559: 399-420.

132. Chu C.J., Djikstra J., Lai M.-Z., Hong K., Szoka F.C. (1990) Efficiency of cytoplasmic delivery by pH-sensitive liposomes to cell in culture. Pharm.Res 7: 824-834.

133. Lee R.J., Wang S., Turk M.J., Low P.S. (1998) The effect of pH and intraliposomal buffer strength on the rate of liposome content release and intracellular drug delivery. Biosci. Rep. 18: 69-78.

134. Subbarao N.K., Parente R.A., Szoka F.C., Nadasdi L., Pongracz K. (1987) pH-dependent bilayer destabilization by an amphipathic peptide. Biochemistry 26: 2964-2972.

135. Connor J., Norley N., Huang L., (1986) Biodistribution of pH-sensitive immunoliposomes Biochim.Biophys. Acta 884: 474-481.

136. Liu D., Huang L. (1989) Small, but not large unilamellar liposomes composed of dioleoylphosphatidylethanolamine and oleic acid can be stabilized by human plasma. Biochemistry 28: 7700-7707.

137. Collins D., Litzinger D.C., Huang L. (1990) Structural and functional comparision of pH-sensitive liposomes composed of phosphatidylrthanolamine and three different diacylsuccinylglycerols Biochim. Biophys. Acta 884: 234-242.

138. Slepushkin V.A., Simoes S., Dazin P., Newman M.S., Guo L.S., Pedroso de Lima M.C., Duzgunes N. (1997) Sterically stabilized pH-sensitive liposomes. Intracellular delivery of aqueous contents and prolonged circulation *in vivo*. J. Biol. Chem. 272: 2382-2388.

139. Hui S.W., Langner M., Zhao Y.L., Ross p., Hurley E., Chan K., (1996) The role of helper lipids in cationic liposome-mediated gene transfer. Biophys J. 72: 590-599.
140. WWW.GENETOOLS,LLC.
141. Summerton J.E. (1999) Morpholino antisense oligomers: the case for an Rhase H-independent structural type. Biochim.Biophys.Acta 1489: 141-158.
142. Morcos P.A. (1999) Gene switching: a strategy for analyzing a broad range of mutations using steric block antisense oligo. Methods in Enzymology 313: 174-189.
143. Lawn R.M. (1999) The Tangier disease gene product ABC1 controls the cellular apolipoprotein-mediated lipid removal pathway. J. Clin. Invest. 8: 104.
144. Giles R.V., Spiller D.G., Clark R.E., Tidd D.M. (1999) Antisense morpholino oligonucleotide analog induces missplicing of C-myc mRNA. Antisense Nucleic Acid. Drug Dev. 9: 213-220.
145. Schmajuk G., Sierakowska H., Kole R. (1999) Antisense oligonucleotides with different backbones. Modification of splicing pathway and efficacy of uptake. J. Biol. Chem. 274: 21783-21789.
146. Summerton J.E. (1997) Morpholino antisense oligomers: design, preparation, and properties. Antisense Nucleic Acid Drug Dev. 7: 187-195.
147. Partridge M., Vincent A., Matthews P., Puma J., Stein D., Summerton J. (1996) A simple method for delivering morpholino antisense oligos into the cytoplasm of cells. Antisense Nucleic Acid Drug Dev. 6: 169-175.
148. Special Issue: Morpholino Gene Knockdowns. (2001) Genesis: 30 (3)- all 27 papers.
149. Draper B.W., Marcos P.A., Kimmel C.B. (2001) Inhibition of zebrafish fgf8 pre-mRNA splicing with Morpholino oligos: a quantifiable method for gene knockdown. Genesis 30: 154-156.
150. Heasman J. (2002) Morpholino oligos: making sense of antisense?. Dev. Biol. 243: 209-214.
151. Mc Manus M.T., Sharp P.A. (2002) Gene silencing in mammals by small interfering RNAs. Nature Rev. Genet. 3: 737-747.
152. Dillin A. (2003) The specifics of small interfering RNA specificity. Proc. Natl. Acad. Sci USA 100: 6289-6291.
153. Tuschl T. (2002) Expanding small RNA interference. Nature Biotechnol. 20: 446-448.
154. Dias N., Stein C.A. (2002) Antisense oligonucleotides: basic concept and mechanism. Mol. Cancer Ther. 1: 347-355.
155. Dove A. (2002) Antisense and sensibility. Nat. Biotechnol. 20: 121-124.
156. Opalińska J.B., Gewirtz A.M. (2002) Nucleic-acid therapeutics: basis principles and recent applications. Natl. Rev. Drug Discovery 1: 503-514.
157. Morris M.C., Chaloin L., Choob M., Archdeacon J., Heitz F., Divita G. (2004) Combination of a new generation of PNAs with a peptide-based carrier enables efficient targeting of cell cycle progression. Gene Ther. 11: 757-764.
158. Sazani P., Vacek M.M., Kole R. (2002) Short-term and long-term modulation of gene expression by antisense therapeutics. Curr. Opin Biotechnol. 13: 468-472.
159. Kurrech J. (2003) Antisense technologies: improvement though novel chemical modifications. Eur. J. Biochem. 270: 1628-1644.
160. Nielsen P.E. (1999) Peptide nucleic acids as therapeutic agents. Curr. Opin Biotechnol. 9: 353-357.
161. Good L., Awasthi S.K., Dryselius R., Larsson O., Nielsen P.E. (2001) Bactericidal antisense effect of peptide-PNA conjugates. Nat. Biotechnol. 19: 360-364.
162. Egholm M., Buchardt O., Christensen L., Behrens C., Freier S.M., Driver D.A., Berg R.H., Kim S.K., Norden B., Nielsen P.E. (1993) PNA hybridizes to complementary oligonucleotides obeying the Watson-Crick hydrogen-bonding rules. Nature 365: 566-568.

163. Nielsen P.E., Egholm M., Berg R.H., Buchardt O. (1991) Sequence-selective recognation of DNA by strand displacement with a thymine-substituted polyamide. Science 254: 1497-1500.

164. Koppelhus U., Nielsen P.E. (2003) Cellular delivery of peptide nucleic acid. Adv. Drug Delivery Rev. 55: 267-280.

165. Aldrian-Herrada G. Desarmenien M.G., Orcel H., Boissin-Agasse L., Mery J., Brugidou J., Rabie A. (1998) A peptide nucleic acids (PNA) is more rapidly internalized in cultured neurons when coupled to a retro-inverso delivery peptide. The antisense activity depresses the target mRNA and protein in magnocellular oxytocin neurons. Nucleic Acids Res. 26: 4910-4916.

166. Braun K., Peschke P., Pipkom R., Lampel S., Wachsmuth M., Waldeck W., Friedrich E., Debus J. (2002) A biological transporter for the delivery of peptide nucleic acids (PNAs) to the nuclear compartment of living cells. J. Mol. Biol. 318: 237-243.

167. Pooga M., Soomets U., Hallbrink M., Valkna A., Saar K., Rezaei K., Kahl U., Hao J.X., Xu X.J., Wiesenfeld-Halin Z., Hokfelt T., Bartfai T., Langel U. (1998) Cell penetrating PNA constructs regulate galanin receptor levels and modify pain transmission *iv vivo*. Nat. Biotechnol. 16: 857-861.

168. Koppelhus U., Awasthi S.K., Zachar V., Holst H.U., Ebbsen P., Nielsen P.E. (2002) Cell-dependent differential cellular uptake of PNA, peptide, and PNA-peptide conjugates. Antisense Nucleic Acids Drug Dev. 2: 51-63.

169. Cutrona G., Carpaneto E.M., Ulivi M., Roncella S., Landt O., Ferrarini M., Boffa L.C. (2000) Effects in live cells of c-myc anti-gene PNA linked to a nuclear localization signal. Nat. Biotechnol. 18: 300-303.

170. Branden L.J., Christensson B., C.I. (2001) *In vivo* nuclear delivery of oligonucleotides via hybridizing bifunctional peptide. Gene Ther. 8: 84-87.

171. Sazani P., Gemignani F., Kang S.H., Maier M.A., Manoharan M., Persmark M., Bortner D., Kole R. (2002) Systematically delivered antisense oligomers upregulate gene expression in mouse tissues. Nat. Biotechnol. 12: 1228-1233.

172. Tackett A., Corey D. (2002) Non-Watson – Crick interactions between PNA and DNA inhibit the ATPase activity of bacteriophage T4 Dda helicase. Nucleic Acids Res. 30: 950-957.

173. Braasch D., Corey D. (2001) Synthesis, analysis, purification, and intracellular delivery of peptide nucleic acids. Methods 23: 97-107.

174. Weiler J., Gausepohl H., Hauser N., Jensen O.N., Hoheisel J.D. (1997) Hybridisation based DNA screening on peptide nucleic acids (PNA) oligomer arrays. Nucleic Acids Res. 25: 2792-2799.

175. Bergman F., Bannwarth W., Tam S. (1995) Solid phase synthesis of directly linked PNA-DNA hybribs. Tetrahedron Lett. 36: 6823-6826.

176. Efimov V.A., Chakhmakhcheva O.G., Archdeacon J., Fernandez J.M., Fedorkin O.N., Dorokhov Y.L., Atabekov J.G. (2001) Detection of 5'-cap structure of messenger RNAs with the use of the cap-jumping approach. Nucleic Acids Res. 29: 4751-4759.

177. Efimov V.A., Choob M., Buryakova A., Phelan D., Chakhmakhcheva O. (2001) PNA-related oligonucleotides mimics and their evaluation for nucleic acids hybridization studies and analysis. Nucleosides Nucleotides Nucleic Acids 20: 419-428.

178. Efimov V.A., Chakhmakhcheva O.G. (1999) Conjugates of polyacrylamide with oligonucleotides and their mimetics for diagnostic purpose. Bioorg Chim. 25: 848-854.

179. Morris M.C., Depollier J., Mery J., Heitz F., Divita G. (2001) A peptide carrier for the delivery of biologically active proteins into mammalians cells. Nat. Biotechnol. 19: 1173-1176.

180. Bordi F., Cametti C. (2002) Salt-induced aggregation in cationic liposome aqueous suspension resulting in multi-step self-assembling complexes. Colloids and Surface B: Biointerfaces 26: 341-350.

181. Eastman S.J., Siegel C., Tousignant J., Smith A.E., Chen S.H., Sheule R.K. (1997) Biophysical characterization of cationic lipid-DNA complex. Biochim. Biophys. Acta 1325: 41-62.

182. Oberle V., Bakowsky U., Zuhorn I.S., Hoekstra D. (2000) Lipoplex formation under equilibrium condition reveals a three-step biophys mechanism. Biophys J. 79: 1447-1454.

183. Verwey E.J.W., Overbeck J.T.G. (1948) Theory of stability liophobic colloids. Elsevier, Amsterdam.

184. Overbeck J.T.G. (1948) in: H.R. Kruyt (Ed.), Colloid Science, Elsevier, Amsterdam.

185. Lasic D.D., (1977) Gene delivery , in: D.D. Lasic (Ed.), Liposomes in Gene Delivery, CRS Press, Boca Raton, FL, pp. 53-66.

186. Stamatatos L., Leventis R., Zuckermann M.J., Silvius I.R. (1988) Interactions of cationic lipid vesicles with negatively charged phospholipids vesicles and biological membranes. Biochemistry 27: 3917-3925.

187. Dan N. (1998) The structure of DNA complexes with cationic liposomes-cylindrical or flat bilayers?. Biochim. Biophys. Acta 1369: 34-38.

188. Dan N. (1996) Formation of ordered domains in membrane-bound DNA. Biophys. J. 71: 1267-1270.

189. Gennis R.B. (1989) Biomembranes-Molecular Structure and function, Springer-Verlag, Berlin.

190. Israelachvili, J. (1992) Intermolecular and Surface Forces, 2nd ed. Academic press, New York.

191. Felgner P.L., Tsai Y.J., Sukhu L., Wheeler C.J., Manthorpe M., Marshal J., Cheng S.H. (1995) Improved cationic lipid formulation for *in vivo* gene therapy. Ann. NY Acad. Sci. 772: 126-139.

192. Behr J.P. (1994) Gene transfer with synthetic cationic amphiphiles: prospect for gene therapy Bioconjugate Chem. 5: 382-389.

193. Niculescu-Duvaz D., Heyes J., Springer C.J. (2003) Structure-activity relationship in cationic lipid mediated gene transfection. Curr. Med. Chem 10:1233-1236.

194. Radler J.O., Koltover I., Salditt T., Safinya C.R. (1997) Structure of DNA-cationic liposomes complexes: DNA intercalation in multilamellar membranes in distinct interhelical packing regimes Science 275: 810-814.

195. Templeton N.S., Lasic D.D., Frederik P.M Strey H., Roberts D.D., Pavlakis G.N. (1997) Improved DNA:liposome complexes for increased systemic delivery and gene expression Nat. Biotechnol. 15: 647-652.

196. May S., Ben-Shaul A. (1997) DNA-lipid complexes: stability of honeycomb-like and spaghetti-like structures. Biophys. J. 73: 2427-2440.

197. Kral T., Hof M., Jurkiewicz P., Laner M. (2002) Fluorescence correlation spectroscopy (FCS) as a tool to study DNA condensation with hexadecyltrimethylammonium bromide (HTAB). Cell Mol. Biol. Lett. 7: 203-211.

198. Bloomfield V.A. (1997) DNA condensation by multivalent cations. Biopolymers 44: 269-282.

199. Kral T., Hof M., Langner M. (2002) The effect of spermine on plasmid condensation and dye release observed by fluorescence correlation spectroscopy. Biol. Chem. 383: 331-335.

200. Jurkiewicz P., Okruszek A., Hof M., Langner M. (2003) Associating oligonucleotides with positively charged liposomes. Cell. Mol. Biol. Lett. 8: 77-84.

Chapter 10

DICATIONIC DEGA-BASED LIPID SYSTEMS FOR GENE TRANSFER AND DELIVERY: SUPRAMOLECULAR STRUCTURE AND ACTIVITY

Ashraf S. Elkady [1]* and Renat I. Zhdanov [2]

[1]*Egyptian Atomic Energy Authority (EAEA), NRC, P.O. Box 13759, Cairo, Egypt;*
[2]*Institute of General pathology and Pathophysiology, Russian Academy of medical Sciences, 125315 Moscow*

Abstract: A novel type of dicationic lipid (DEGA) based on L-Glutamic acid, with different spacer arms between the two polar heads of lipid chain groups, was recently synthesised. The potential of cationic liposomes, prepared from DEGA, for gene delivery was tested by monitoring gene expression in human cancerous cells. Besides, Coherent Phase Microscopy was applied to study the DNA-DEGA complex formation. Encapsulation of DNA molecules by the cationic liposomes, and a significant increase in the optical density and phase height were observed for the complex, compared to individual liposomes. The incubation time was found to play an important role in the condensation and/or encapsulation of DNA molecules by the cationic lipid. Moreover, the conformational and structural peculiarities of DNA-DEGA complexes were studied using Atomic Force Microscopy (AFM). It is shown that short spacer DEGA liposomes form morphologically more compact particles with plasmid DNA of spherical and toroidal form (50-120nm) and effectively transfer functional genes in eukaryotic cellular lines; while those with long spacers form unstable amorphous aggregates and were not potent for cell transfection.

Key words: AFM, DNA-lipid complex, nanomedicine, supramolecular complex, transfection of human cell line.

---

* Corresponding author: Dr. A.S. Elkady, email: ashraf_elkady@yahoo.com

M.R. Mozafari (ed.), Nanocarrier Technologies: Frontiers of Nanotherapy, 175–190.

1. INTRODUCTION

The 21st century has witnessed a rapid and continuing development in the field of nanotechnology that allowed for applications well beyond the scope of physics, chemistry and biology. In order to face major challenges, this science should combine its efforts with many other sciences, e.g. molecular biology, biotechnology and nanotherapy. Applications of nanotechnology for treatment, diagnosis, monitoring, and control of biological systems has recently been referred to as "nanomedicine" by the National Institutes of Health. Research into the rational delivery and targeting of pharmaceutical, therapeutic, and diagnostic agents is at the forefront of projects in nanomedicine. These involve the identification of precise targets (cells and receptors) related to specific clinical conditions and choice of the appropriate nanocarriers to achieve the required responses while minimizing the side effects [1]. Of special interest is the process of molecular self-assembly, in which molecules or parts of molecules spontaneously form ordered aggregates without any human intervention, thus introducing a 'bottom-up' approach to the fabrication of objects specified with nanometre precision. The molecular structures and intermolecular interactions of DNA are particularly amenable to the design and synthesis of complex molecular objects [2]. Particularly, the self-assembly of amphiphilic molecules constitute one of the most fundamental mechanisms for the construction of soft condensed matter biomaterials [3]. It is also well known that lipids, as well as mixtures of anionic and cationic single chain surfactants, can readily form bilayers [4, 5] that can adopt a variety of distinct geometric forms: they can fold into soft vesicles or random bilayers (the so-called sponge phase) or form ordered stacks of flat or undulating membranes [6].

In the present work we investigated lipoplexes formed from DNA and recently synthesized dicationic lipid analogues [7-9], DEGA [(N'-dimethyl-N''-dimethyl)-bis-(dihexadecylglutamate) butan-L-glutamic acid derivatives] (or heptane- L-glutamic acid derivatives), and their applications *in vitro* transfection and gene expression in human cancerous cells. We emphasized on the physico-chemical and biophysical characterization of the self-assembled lipoplexes, and their structure-activity relationships. The article will highlight some of the key parameters that are crucial for optimising gene transfer by correlating the structural features of nanoscale vehicles to transfection efficiencies.

2.   EXPERIMENTAL

A novel type of dicationic lipid based on L-glutamic acid, a group of original lipopeptides with two quarternary ammonium polar heads was synthesized [7]. The molecule of dimer: $[CH_3(CH_2)_{15}O]_2Glu(CH_3)_2N^+-$ $(CH_2)_n-N^+(CH_3)_2Glu[O(CH_2)_{15}CH_3]_2$ is based on hydrophobic block (four saturated hydrocarbon chains, corresponding residues of hexadecyl alcohols) and a complex polar group modified aminoacids containing two quarternary ammonium salts linked through hydrophobic aliphatic spacer arms with different length alkane (n = 3-11).

Cationic liposomes were prepared using lipid hydration method for different lipid derivatives [10]. Ten milligrams of lipid dimer was dissolved in 1ml chloroform. The suspension solution was placed in a rotor vacuum system to evaporate the organic solvent until the sample is completely dry. The dry lipid film was suspended in 1ml of NaCl aqueous buffer (0.89%). The dispersion was heated to $60°C$, and vortexed producing large multilamellar vesicles (LMVs). The vesicle suspension was sonicated to clarity for 10 min. The liposome preparation was then extruded by using LiposoFast apparatus (Avestin, Canada) equiped with 100nm pore cellulose membrane filter (Millipore, U.S.A). After 25 passages through double 100nm filters, clean mixtures of small unilamellar vesicles were obtained. The vesicles were stored at $4°C$ in a nitrogen atmosphere and used within a week. Lipoplexes were prepared by mixing aqueous plasmid DNA pEGFP-PN1 (Clontech, 4.7 kb) with DEGA liposomes or lipofectin® (Invitrogen Corporation) at desired charge ratios. They were used in measurements after incubation for 2 h and longer.

Laser phase microscopy of liposomes or their complex with DNA (lipoplex) was performed with an Airyscan computer phase instrument [11, 12]. A specimen was placed in a cuvette (a polished silicon wafer capped with a cover slip) on the stage of a modified Linnik microinterferometer with a He–Ne laser as a light source. Linear-periodic modulation of the reference wave phase was achieved by the use of a mirror with a piezoelectric modulator. A dissector consisting of a coordinate-sensitive LI-620 photo-detector and an electronic module recorded the interferometer signal and converted the analog signal to digital local phase data. Digital phase images and local fluctuations of the optical path were analyzed using an original special-purpose computer program.

Atomic Force Microscopy (AFM) for DNA-DEGA lipoplexes was done using NanoScope III-a (Digital Instruments, Santa Barbara, CA) equiped with commercial cantilivers with oxide sharpened silicon nitride (spring constant 0.32 N/m) in contact mode, and etched silicon with resonance frequency 200-400 KHz in the measurements carried out in tapping mode. The scan frequency in contact mode was 5Hz, and in tapping mode was 1Hz. Freshly cleaved mica was used as substrate onto which 5µl smaple was placed and dried in air after excess solution was drained by filter paper.

Fluorescence Microscopy was carried out to test the lipofection efficiency, using Carl Zeiss AXIOSKOP-20 fluorescent microscope (Jena, Germany), equipped with numerical system of image analysis and a system of passing and blocking light filters FITC/TRITC for double fluorescent labeling. Taking statical photoimages and subsequent image treatment are carried out using numerical system based on colour integrating 3-matrice (3CCD) videocamera "Sony" and station for non-linear videomontage (P4, 1.5 GGz, FRGR Matrox Meteor). KS100 (Carl Zeiss, Jena, Germany), Adobe Photoshop, and Ulead Media Studio programs are used for the above purpose.

Cell culture and lipofection assays were done as follows: MCF, 239 and SKOV–3 cells (ATCC) were seeded with density of 2.10^5 per well of a 96–well tissue culture plates in 200µl of the DMEM growth medium (Gibco^{TM}), supplemented with 10% serum. The cells were then incubated at 37°C in a CO_2 incubator for 24 hours. A diluted pDNA pEGFP-N1 (Clontech), 2µg per well, was mixed gently with the dicationic liposomes or lipofectin® (Invitrogen Corporation) with different weight ratios (r) in serum-free growth medium DMEM. The mixture was incubated at room temperature for 2h. The cells were then washed once with serum-free growth medium without antibacterial agents, overlayed with DNA–CL complexes and incubated for 24h at 37°C in a CO_2 incubator. Afterwards, the DNA containing medium was replaced with growth medium supplemented with the normal percentage of serum and the cells were incubated at 37°C in a CO_2 incubator for another 24h [10].

Transfection efficiency was determined using fluorescent microscopy by counting coloured cells 24–48 h after lipofection following an expression of reporter genes. Transfection efficiency is counted as a part of cells, which express fluorescent protein, divided by total cell number (in percent).

Small angle neutron scattering (SANS) was carried out using D22 facility, Grenoble, France. D22 is the small angle neutron scattering facility with the highest flux at the sample position, in a wavelength range of 4 to 40 Å. It has a large dynamic range of momentum transfer, i.e. from 8×10^{-3} to 0.6 Å^{-1} (10 to 6000 Å in the direct space). A relative narrow wavelength band (fwhm $\Delta\lambda/\lambda = 10\%$) is chosen by a DORNIER velocity selector from the neutrons leaving the horizontal cold source. The selector has a 25cm long rotating drum with helical lamellae shaped in a three-dimensional mould. The maximal speed is 28300 rpm filtering a wavelength of 4.6 Å. The selector can be rotated around its vertical axis for tuning the resolution and/or reaching shorter wavelengths. D22 possesses the largest area multidetector (^{3}He) of all small-angle scattering instruments (active area 96×96 cm^2), with a pixel size of 7.5×7.5 mm^2. The detector efficiency is wavelength dependant and equals to 80%, for $\lambda = 6$ Å. Standard sample holders of various Hellma cells contained in thermostatted sample environment with a circulating bath (from –5 to 85°C), were used during measurements [14].

3. RESULTS AND DISCUSSIONS

The affinity of free DEGA liposomes to cell membrane was tested by using the lipophilic fluorescent dye, cumarin, and fluorescence microscopy. A high affinity to cell membrane and subsequent (within 15 minutes) accumulation in cytoplasmic compartments are evident (see Fig. 10-1 a, b). The entrapment of liposomes by cells could be due to electrostatic interaction with the negatively charged cell membrane.

To assess the relevance of the dicationic lipid DEGA for gene delivery, we performed cell transfection experiments. For gene transfection into eukaryotic cells, we used lipid DEGA as a liposomal preparation in case of DEGA-3 and DEGA-4 and a mixture of micellar and vesicular in case of DEGA-7. Transfection conditions were identical to those determined to be optimal for mammalian cell lines, i.e. conditions characterized by DNA-lipid aggregates bearing a strongly positive mean charge (r = 6-10) [15].

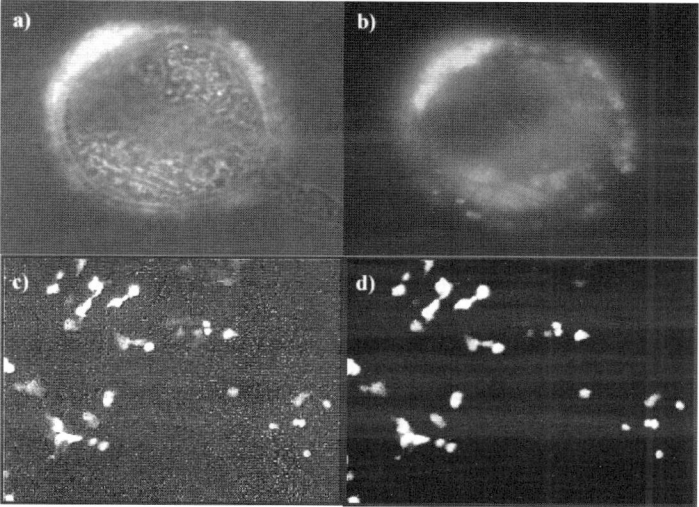

Figure 10-1. (a, b) Interaction of cumarin-labelled DEGA-3 liposomes with 293 cell line. The accumulation of fluorescence in cytoplasmic compartments took place in 15 min. (c, d) DEGA-3 mediated gene transfection with pEGFP-N1-DEGA-3 lipoplex: transfected cells emit green fluorescence: a, c) FM images taken in bright field, b, d) FM images taken in dark field.

The potential of DEGA liposomes for delivering functional (reporter) genes into eukaryotic cells was examined by monitoring gene expression in different types of established cell lines, e.g. SKOV-3, 293 and MCF-7 [8-10]. Besides, for comparison purposes, the cell lines used were also transfected with lipofectin[R], a commercially available cationic lipid that is very efficient for transfering genes into eukaryotic cells *in vitro*. Representative example for 239 cell lines transfection is given in Fig. 10-1 (b, c).

Surprisingly, the cationic lipids varied in their ability to mediate cell transfection: lipoplexes composed of DEGA-3 and DEGA-4 (i.e. with short spacers between cationic moieties) showed relatively high transfection efficiencies (Fig. 10-1 a, b), compared to those composed from DEGA-7 and DEGA-11 (with longer spacers). For instance, the efficiency of gene transfer in 293 cells, using lipoplexes based on positively charged DNA aggregates with DEGA-3 liposomes was 20% ± 2%, while for DEGA-7 and DEGA-11 did not exceed 1-5%. The efficiency of DEGA-3 is comparable with gene transfer efficiency for lipofectin[R] 25% ± 2% for the same cell lines. Thus, our data demonstrate that DEGA, a dicationic lipid based on glutamic acid derivatives, is an efficient reagent for transfection of various cell types, if used with short aliphatic spacer between the 2 cationic moieties. Remarkably, though we haven't used a helper lipid like DOPE in DEGA preparations, they showed satisfactory transfection efficiencies.

In addition, it is generally agreed that lipoplexes are internalized by means of nonspecific endocytosis after binding of positively charged lipoplexes to anionic residues on the cell surface [16-19]. Thus, detection of typically expressed genes in cell lines transfected with positively charged DEGA-DNA aggregates is consistent with electrostatic-driven endocytosis being the major route of DNA uptake during DEGA-mediated transfection. Besides, as endocytosis leaves the DNA still two membranes away from the nucleus, escape from the endosome and release into the cytosol is a highly critical step that needs to be carried out by the lipoplexes [20]. It has been proposed recently that destabilization of the endosomal membrane may induce the flip-flop of anionic lipids from the cytoplasm-facing monolayer of the endosome, leading thereby to the formation of a charge-neutral ion pair with the cationic lipid and to subsequent dissociation of the DNA from the complex and its release into the cytoplasm [21].

Further, it was important to elucidate the structural and morphological peculiarities of DEGA-lipoplexes in order to understand their different transfection efficiencies (activities). Therefore, we invistigated some of the structural and interfacial features for DNA-DEGA complexes from micro-down to nanometer lenght scale, using coherent phase and atomic force microscopy respectively, combined with small angle neutron scattering.

Topograms and 3D phase height images for unextruded liposomes, formed from lipid DEGA-4, and their complex with pDNA (pEGFP-N1) as viewed with CPM are displayed in Fig. 10-2. It's noticeable that a higher contrast due to increasing in optical density is obtained for DNA-DEGA-4 complexes, compared to free liposomes (cf. Figs. 10-2a and 10-2c). In contrast to free liposome solution where no objects >0.3μm were found, globular vesicles ≥1μm were observed for their DNA complexes. The formed complexes (lipoplexes) have either spherical or ellipsoidal forms and look like vesicles that have entrapped dense material of tightly bound DNA. This finding is consistent with Cryo-EM images reported for genosomes prepared from large DOTAP and DOTAP-chol and DOTAP: DOPE vesicles [22]. Besides, the formation of dense birefringent condensed globules ~1μm were directly observed for another kind of lipoplexes composed from DOPC/DOTAP and λ-DNA, using differential interference and polarized light microscopy [23].

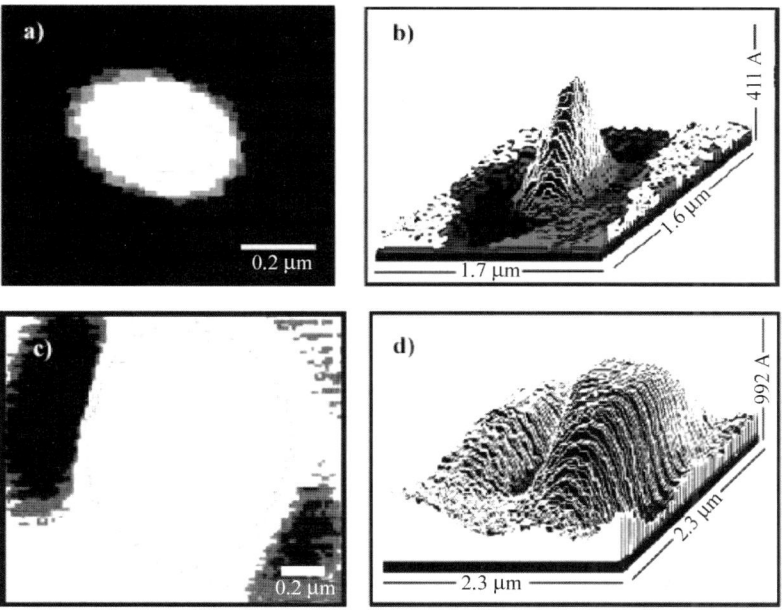

Figure 10-2. Phase images for DEGA-4 liposomes (a, b) and their complex with pDNA (c, d), illustrating the significant increase in phase height upon complex formation.

Based on the analysis of CPM images, one can estimate that unextruded liposomes from lipid DEGA-4 have the average cross-section 225 ± 7nm, while the cross-section of their aggregates with DNA is around 1000 ± 15nm. Figs. 10-2b and 10-2d represent the corresponding 3D images for the 2D topograms in Figs. 10-2a and 10-2c. From Fig. 10-2b it follows that liposomes typically have a narrow distribution of phase height (h) corresponding to average cross-section size $225 \times 255 \times 40$nm. The 3D topograms for aggregate unit structures has a much bigger average size $1000 \times 1380 \times 95$ nm (see Fig. 10-2d for example). Thus on semi-macroscopic length scale, the addition of DNA molecules to DEGA-4 liposomes induces a topological transition from liposomes into collapsed optically dense condensates in the form of spherical and ellipsoidal (bean-like) vesicles with sizes on the order of $1 \mu m$.

The complex (lipoplex) images were obtained after 2 hours of incubation time. Shorter times of incubation resulted in complexes of smaller size: a result that testified the importance of the time of incubation for complex formation between DNA molecules and the cationic lipid, and formation of optically dense condensed particles. The low optical density and contrast at shorter times didn't enable us to monitor early kinematical changes in complex formation.

The data on phase height distribution, obtained by measuring parameters of many particles, lead to a statistically reliable conclusion that a significant increase in the average sizes of DNA-DEGA-4 aggregates occur compared to those for free liposomes. The average sizes for aggregates ($<\varnothing> = 425$nm, $<h> = 60.5$nm) compared to free liposomes ($<\varnothing> = 260$, $<h> = 18$ nm), show the existence of two particle populations with considerably differing size parameters. The directly observed increase in optical density and phase height distribution for aggregates, compared to free liposomes could be attributed to encapsulation of DNA molecules by DEGA-4 liposomes.

Further, we used atomic force microscopy (AFM) for characterizing the size and morphology of complexes formed from pDNA and extruded DEGA liposomes on nanoscale. A series of AFM images for pDNA-DEGA-4 mixtures as a function of lipid to DNA weight ratio (r) are represented in Fig. 10-3. Depending on the ratio "r", we observed different aggregate structures for lipoplexes after adhesion to the mica surface. At low lipid concentration (r < 3, Fig. 10-3a), beads-on-strings structures are visualized. The complexes exist with excess DNA lipid-coated fibres with distinguishable attached liposomes. As more lipid was added, the aggregates formed compact particles of toroidal shape (r = 7.5, $<\varnothing> = 120$ nm, Fig. 10-3b). These toroidal condensates transformed to globular structures of smaller size at a higher lipid concentration (r = 0, $\varnothing = 0$-65nm). Indeed, the toroidal-globular transition was hetherto monitored at high concentrations of cationic surfactant [24].

Moreover, it was noted that the number of toroidal and globular condensates is proportional to the incubation time of the complexes, i.e longer incubation time results in an increase of the DNA fraction condensed by the cationic surfactant and formation of maturated nanoparticles. Thus the incubation time could be considered as an important kinematical parameter controlling the DNA condensation and/or encapsulation process. This conclusion is consistent with the observation of intermediate morphologies at shorter incubation times (data not shown), and with results obtained in synchrotron studies, which indicate that the incubation time removes the DNA positional disorder in the DNA one-dimensional lattice between lipid bilayers [25].

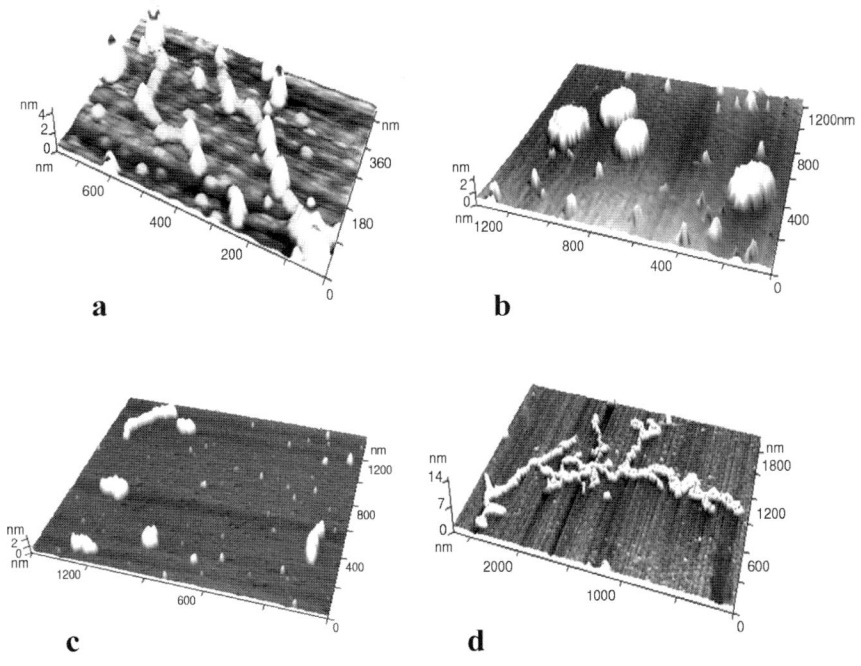

Figure 10-3. AFM images of pDNA-DEGA-4 (a, b) and pDNA-DEGA-7 (c, d) complexes after adhesion to mica surface: a) beads-on-strings structure, r (lipid:DNA weight ratio) = 1.5; b) toroidal structure, r = 7.5; c) mixture of rods and deformed toroids obtained at r = 6, incubation time one hour; d) clustering of lipoplex particles after 2 hours of incubation.

The reason that liposomes and lipoplexes visualized by CPM are larger in size compared to those visualized by AFM, relies on the fact that liposomes used in CPM measurements were slightly sonicated and not extruded, allowing their size to be within the resolution limit of such optical technique. However, it's expected that if liposomes with excess surface area are used, complete encapsulation of DNA can occur [22]. It's clearly seen from CPM images that lipoplexes prepared from large DEGA-4 vesicles (diameter 250nm) are somehow different. They look like vesicles that have entrapped dense material of DNA.

Interestingly, differences were observed in the aggregation behavior for DEGA-7 lipoplexes: a mixture of rod-like and deformed toroids were observed at $r = 6$ (see Fig. 10-3c). These different structural units could be due to the fact that DEGA-7 suspension contains both of micelles and large vesicles and upon interaction with DNA, they give rise to different structures. However, these structural units were found to be unstable in physiological solution and aggregate within 2 hours, forming larger structures (Fig. 10-3d). Thus, while DEGA-3 and DEGA-4 lipoplexes, exhibiting high transfection activity form distinct compact condensed particles in toroid and globule forms, which are stable against aggregation; the formation of multimolecular clusters (~3μm) and amorphous aggregates have been noted for lipids with longer spacers, DEGA-7 and DEGA-11. No compact condensates or complete DNA encapsualtion were observed with longer sapcer arms, i.e. DEGA-22 lipoplexes [26]. An intriguing aspect for the latter lipids (with long spacer arm) is that part of the initially formed lipoplexes in bulk phase are unstanble and eventually aggregate and/or merge further into larger complexes within 2 hours of incubation. The size of these complexes exceeds that of plasmid and may acquire diameters in the order of microns. It is more likely that this fraction will be excluded from involvement in cellular transfection, as such a size will prevent cellular uptake. Presumably, as suggested by others for another kind of amphiphiles [27], this fraction may actually originate from vesicles that in the early phase of preparation still display surface-bound plasmid, or at least uncovered parts of DNA. It also implies that plasmids that are rapidly and completely covered with lipid (as the case with DEGA-3 and DEGA-4) will remain in bulk solution as stable particles with little tendency to cluster, and capable of bringing about efficient transfection. It's therefore possible, that the stability of lipoplexes may depend on the amphiphile-dependent packing efficiency of plasmids, implying a degree of flexibility that would ensure rapid and complete shielding of DNA molecules, thereby preventing unlimited growth of complexes, which would hamper their internalization. Thus, the compact, moderate size (50-120nm) and stability of DEGA-3 and DEGA-4 lipoplexes, may justify their superior transfection efficiency over the other derivatives, i.e. DEGA-7 and DEGA-11.

For further characterization of the solution structural features DNA-DEGA complexes, we also used small angle neutron scattering measurements (SANS), iLL, Grenoble. In Fig. 10-4, the scattering intensity of DNA-DEGA-4 lipoplexes is plotted as a function of the neutron scattering vector. The observed Bragg peak at q position ~0.09 $\mathring{A}^{-1}$ revealed a lamellar organization with a regular spacing of 7 ± 0.2nm. In control experiments, neither naked pDNA nor unreacted unilamellar liposomes exhibited diffraction maxima (data not shown), which justified that the lamellar organisation is induced by lipoplex formation.

Thus, the scattering pattern reflects the periodic spacing of lamellar DNA-DEGA lipoplexes. The periodicity of 7nm being consistent with stacks of alternating lipid bilayers with intercalated DNA monolayer. With $\delta = \delta_m + \delta_{D+D2O}$, where δ_m is the thickness of the lipid bilayer and δ_{D+D2O} is the thickness of the DNA double strand and one layer of heavy water ($\delta_{D+D2O} = 2.5$nm), then the thickness of the lipid bilayer would be 4.5nm. As second-order reflection was absent, the lamellar organization presumably does not extend over large distances. Measurement of the half-width of the first-order diffraction peak at 7nm indicated an effective domain size of 27nm, corresponding to about four periodicities and consistent with AFM images (see inset in Fig. 10-4).

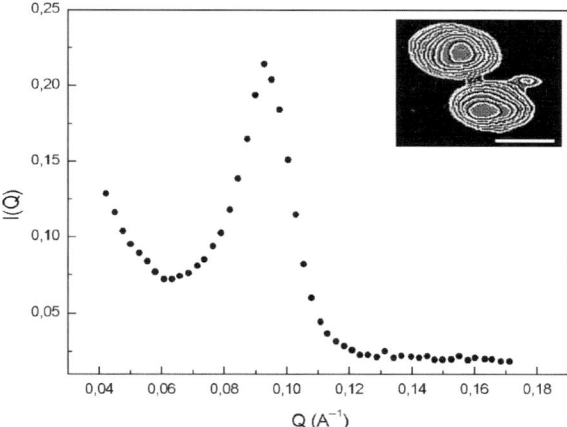

Figure 10-4. Small angle neutron scattering (SANS) scattering intensity of the DNA-DEGA-4 complex at 25°C as a function of the scattering vector (Q). The inset is the corresponding AFM image for the complex, the bar represents 100 nm.

In fact, lamellar phases have been reported for polypeptides complexed with oppositely charged lipids or bilayer-forming surfactants [28-31]. Compared to polypeptides or synthetic polymers, DNA is characterized by greater values of persistence length (50nm) and high charge density. The common feature of these polyelectrolyte lipid composite phases is the self-assembling mechanism: the hydrophobic interaction forming the planar lipid bilayers and a strong electrostatic interaction that is responsible for the alternating layering and in-plane ordering of the cationic lipid and anionic polymer [32].

Besides, the neutron scattering data is consistent with recent synchrotron radiation measurements, indicating a complete topological rearrangement of liposomes and DNA into a multilayer structure with DNA intercalated between the bilayers [23]. It is interesting to note the absence of any Bragg peaks for free liposomes and free DNA at the applied concentrations. Indeed in the absence of DNA, the lipid membranes are expected to exhibit strong long-range interlayer electrostatic repulsions that overwhelm the van der Waals attraction [33, 34]. The DNA that condenses on the dicationic lipid strongly screens the electrostatic interaction between lipid bilayers and leads to condensed multilayers. The average thickness of the heavy water gap δ_{D+D2O} = 2.5nm is just sufficient to accommodate one monolayer of B-DNA (diameter 2nm) including a hydration shell [35].

Moreover, the periodicity of around 7nm indirectly tells us that charges are not dehydrated. If this is the case, DNA would contribute around 2nm, lipid bilayers would tilt, and overall periodicity of 4.7 to 5nm would be observed. This observation may also indicate that fluid lipid bilayers are important for transfection [22].

Thus, for one type of cationic lipid (DEGA-4), we observed 4 characteristic structures, depending on DNA/DCL ratio (r): «beads-on-strings» (r < 3, Fig. 10-3a), lamellar (r ~ 3-7, Fig. 10-4), toroidal and globular (r = 6 or r = 10, Fig. 10-3b). These types of condensates correspend to diffrent conformations due to changes in secondary and tertiary structure of DNA molecules.

Therefore, short spacer DEGA lipoplexes, as visualized by AFM, exhibited various ordered patterns of supramolecular organization upon interaction with DNA, depending on the lipid/DNA charge ratio. The condensation and/or encapsulation density of DNA in cationic complex (in forms of toroids and globules) is higher than in anionic complex (in forms of beads-on strings and spagetti-like structures). The complexes near to the isoneutral point showed instability in their aggregation size. Highly effective complexes for gene transfer were obtained for cationic complexes (r ≥ 6). It seems that the high packing density for plasmids when they are entrapped either in multilamellar complex or condensed in toroidal forms allows for a larger number of protected plasmids to reach the cell nucleus and express the required gene. The size dependence of the aggregates can be understood in terms of a charge-stabilized colloidal suspension. As described by the well known DLVO theory, the competition of repulsive electrostatic and attractive van der Waals forces determines if a colloidal suspension is stable or tends to precipitate [36, 37].

4.   CONCLUSIONS

To conclude, the structure-activity relationship studies carried out for lipoplexes prepared from a series of newly synthesized dicationic lipids DEGA showed that the stability of DEGA-DNA complex unit structure is an important consideration in the design or selection of cationic lipid for self-assembling DNA delivery systems that could be used in nanotherapy. It is shown, that DEGA liposomes with short spacers form morphologically more compact particles of the globular and toroidal forms (50-120nm), which are stable in solution (showed minimal aggregation of their DNA complexes) and effectively transfer functional genes to cultured cellular lines. On the other hand, complexes formed from long spacer liposomes form the big, noncompact amorphous aggregates, which are unstable and do not transfer genes in cultured cells.

Acknowledgements
We express our gratitude to A.R. Khoklhlov, I.V. Yaminsky, V.P. Tychinsky, Yu.L. Sebyakin, A.A. Moskovtsev and Peter Timmins for their help in conducting experiments and discussion of the results.

5.   REFERENCES

1. Moghimi, S.M., Hunter, A.C. and Murray, J.C. (2005) Nanomedicine: current status and future prospects, The FASEB Journal, 19: 311-330.
2. Winfree, E., Furong, L., Wenzler, L. and Seeman, N.C. (1998) Design and self-assembly of two-dimensional DNA crystals. Nature, 394: 539-544.
3. Wong, G.C.L., Tang, J.X., Lin, A., Li, Y., Janmey, P.A. and Safinya, C.R. (2000) Hierarchical self-assembly of F-Actin cationic lipid complexes: stacked three-layer tubule networks. Science, 288: 2035-2039.
4. Sackmann, E. and Lipowsky, R. (1995) Handbook of Biological Physics Vol. 1B (Hoff, A.J. ed.) North-Holland, Amsterdam.

5. Kaler, E.W., Murthy, A.K., Rodriguez, B.E. and Zasadzinski, J.A.N. (1989) Spontaneous vesicle formation in aqueous mixtures of single-tailed surfactants. Science, 245, 1371-1374.

6. Dubois, M. and Zemb, T. (1997) Swelling limits for bilayer microstructures: the implosion of lamellar structure versus disordered lamellae. Curr. Opin. Colloid Interf. Sci. 5: 27-37.

7. Polyakova, A.A., Panchenkova, I.A., Skripnikova, M.A., and Sebyakin, Yu.L. (2003) Biol. Membr., 20 (2): 178-183.

8. Elkady, A.S., Tychinskii, V.P., Sebyakin, Y.L., Moskovetsev, A., Zhdanov, R. and Khokhlov, A.R. (2003) On the Structural-Gene Transfer Activity Relationship of DNA Dicationic Lipid Genosomes. Journal Biomolecular Structure and Dynamics, 20 (6): p. 898.

9. Elkady, A.S., Sebyakin, Y., Gallyamov, M., Moskovtsev, A., Bischoff, G., Shmyrina, A., Zhdanov, R. and Khokhlov, A. (2004) DNA-lipid supramolecular complexes: structural and functional peculiarities as studied by scanning atomic force microscopy. In: Micro- and Nanostructures of Biological Systems (G. Bischoff and H.J. Hein, eds.), 2nd edition, Shaker Verlag, Aachen, 32-48.

10. Zhdanov, R.I., Bogdanenko, E.V., Moskovtsev, A.A., Podobed, O.V. and Duzgunes, N. (2003) Liposomes in Gene Delivery and Gene Therapy. Methods in Enzymol. 373: 433-465.

11. Tychinsky, V.P., Kufal, G.E. and Vyshenskaja, T.V. (1997) Measurements of submicron structures with the Airyscan laser phase microscope. Quant. Electronics, 27: 735-739.

12. Tychinsky, V.P. (2001) Coherent Phase Microscopy of intracellular processes. Usp. Fiz. Nauk, 171: 649-662.

13. Filonov, A.S., Gavrilko, D.Y. and Yaminsky, I.V. (2001) FemtoScan SPM Image Processing Software Manual, Advanced Technologies Center, Moscow – 66 p. (http://www.nanoscopy.net).

14. Roland, M. (2003) D22 Manual, located at website: http://whisky.ill.fr/YellowBook/ D22/D22_info/html/D22Manual.html.

15. Aissaoui, A., Oudrhiri, N., Petit, L., Hauchecorne, M., Kan, E., Sainlos, M., Julian, S., Navarro, J., Vigneron, J.P., Lehn J.M. and Lehn, P. (2002) Progress in gene delivery by cationic lipids: guanidinium-cholesterol-based systems as an example. Current Drug Targets, 3: 1-16.

16. Zabner, J., Fasbender, A.J., Moninger, T., Poellinger, K.A. and Welsh, M.J. (1995) Cellular and Molecular Barriers to Gene Transfer by a Cationic Lipid. J. Biol. Chem. 270: 18997-19007.

17. Labat-Moleur, F., Steffan, A.M., Brisson, C., Perron, H., Feugeas, O., Furstenberger, P., Oberling, F., Brambilla, E. and Behr, J.P. (1996) An electron microscopy study into the mechanism of gene transfer with lipopolyamines. Gene Ther. 3: 1010-1017.

18. Wrobel, I. and Collins, D. (1995) Fusion of cationic liposomes with mammalian cells occurs after endocytosis. Biochim. Biophys. Acta, 1235: 296-304.

19. Zhou, X. and Huang, L. (1994) DNA transfection mediated by cationic liposomes containing lipopolylysine: characterization and mechanism of action. Biochim. Biophys. Acta, 1189: 195-203.

20. Pitard, B., Oudrhiri, N., Vigneron, J.P., Hauchecorne, M., Aguerre, O., Toury, R., Airiau, M., Ramasawmy, R., Scherman, D., Crouzeti, J., Lehn, J.M. and Lehn, P. (1999) Structural characteristics of supramolecular assemblies formed by guanidinium-cholesterol reagents for gene transfection. Proc. Nat. Acad. Sci. USA, 96: 2621-2626.

21. Xu, Y. and Szoka, F. C. (1996) Mechanism of DNA release from cationic liposome/DNA complexes used in cell transfection. Biochemistry, 35: 5616-5623.

22. Lasic, D. (1997) Liposomes in Gene Delivery, CRC Press, New York.

23. Radler, J.O., Koltover, I., Salditt, T. and Safinya, C.R. (1997) Structure of DNA–Cationic Liposome Complexes: DNA Intercalation in Multilamellar Membranes in Distinct Interhelical Packing Regimes. Science, 275 (5301): 810-814.

24. Vasilevskaya, V.V., Khokhlov, A.R., Kidoaki, S. and Yoshikawa, K. (1997) Structure of collapsed persistent macromolecules: toroid vs. spherical globule. Biopolymers. 41: 51-60.

25. Uhrikova, D., Rapp, G. and Balgavy, P. (2002) Condensed lamellar phase in ternary DNA-DLPC-cationic gemini surfactant system: a small angle synchrotron x-ray diffraction study. Bioelectrochem. 58: 87-95.

26. Elkady, A.S., Sebyakin, Y., Gallyamov, M., Moskovetsov, A., Alexeev, A., Bischoff, G., Skripnikova, M.A., Zhdanov, R. and Khokhlov, A.R. (2002) DNA Supramolecular Complexes: Structural and Functional Peculiarities as Studied by Scanning Atomic Force Microscopy. Conference paper, proceedings of the international conference on Scanning Probe Microscopy SPM-2002, Nizhny Novgorod, March 3-6, 2002, pp. 181-184.

27. Oberle, V., Bakowsky, U., Zuhorn, I.S. and Hoekstra, D. (2000) Lipoplex Formation under Equilibrium Conditions Reveals a Three-Step Mechanism. Biophysical Journal, 79 (3): 1447-1454.

28. de Kruijff, B., Rietveld, A., Telders, N. and Vaandrager, B. (1985) Molecular aspects of the bilayer stabilization induced by poly (L-lysines) of varying size in cardiolipin liposomes. Biochim. Biophys. Acta, 820, 295-304.

29. Antonietti, M., Conrad, J. and Thunemann, A. (1994) Polyelectrolyte-surfactant complexes: A new type of solid, mesomorphous material. Macromolecules, 27, 6007-6011.

30. Ponomarenko, E.A., Waddon, A.J., Bakeev, K.N., Tirrell, D.A. and MacKnight, W.J. (1996) Self-Assembled Complexes of Synthetic Polypeptides and Oppositely Charged Low Molecular Weight Surfactants. Solid-State Properties. Macromolecules, 29: 4340-4345.

31. Zschornig, O., Arnold, K., Richter, W. and Ohki, S. (1992) Dextran sulphate-dependent fusion of liposomes containing cationic stearylamine. Chem. Phys. Lipids, 63: 15-22.

32. Radler, J.O., Koltover, I., Jamieson, A., Salditt, T., and Safinya, C.R. (1998) Structure and Interfacial Aspects of Self-Assembled Cationic Lipid-DNA Gene Carrier Complexes. Langmuir, 14, 4272-4283.

33. Roux, D. and Safinya, C.R. (1988) A synchrotron x-ray study of competing undulation and electrostatic interlayer interactions in fluid multimembrane lyotropic phases. J. Phys. (France) 49, 307.

34. Safinya, C.R. (1986) Phase Transitions in Soft Condensed Matter, Tormod R. and Sherrington D., Eds. (Plenum, New York, 1989), pp. 249-270. Sci. USA. 83, 7132.

35. Podgornik, R., Rau, D.C. and Parsegian, V.A. (1989) The action of interhelical forces on the organization of DNA double helices: Fluctuation-enhanced decay of electrostatic double layer and hydration forces. Macromolecules 22, 1780.

36. Hiemenz, PC. (1986) Principles of colloids and surface chemistry, 2nd edn. Marcer Dekker: New York.

37. Tang, M.X. and Szoka, F.C. (1997) The influence of polymer structure on interactions of cationic polymers with DNA and morphology of the resulting complexes. Gene Therapy, 4: 823-832.

Chapter 11

THE ROLE OF LIPOSOMAL ANTIOXIDANTS IN OXIDATIVE STRESS

Zacharias E. Suntres [1] and Abdelwahab Omri [2]

[1]*Medical Sciences Division, Northern Ontario School of Medicine, Lakehead University, Thunder Bay, Ontario, P7B 5E1;* [2]*The Novel Drug and Vaccine Delivery Systems Facility, Department of Chemistry and Biochemistry, Laurentian University, Sudbury, Ontario, P3E 2C6, Canada*

Abstract: Oxidative stress has been characterized by an elevation in the steady-state concentration of reactive oxygen species including superoxide anion, hydrogen peroxide, and hydroxyl radical. There is increasing evidence connecting oxidative stress with a variety of pathological conditions including cancer, cardiovascular diseases, chronic inflammatory disease, post-ischaemic organ injury, diabetes mellitus, xenobiotic/drug toxicity, and rheumatoid arthritis. A pharmacological strategy in preventing or treating oxidant-induced damage is by administering appropriate antioxidants. Evidence from several studies have shown that administration of antioxidants did not seriously modify the injurious actions of oxidants, an observation attributed mostly to their inability to cross cell membrane barriers and/or to their rapid clearance from cells. Recent advances in controlled delivery systems for drugs such as liposomes have generated an interest in their potential application for the prophylaxis or treatment of oxidant-induced injuries. The relative ease in incorporating hydrophilic and lipophilic therapeutic agents in liposomes; the properties of liposomes to protect therapeutic agents from inactivation; the possibility of directly delivering them to an accessible body site; and their relative non-immunogenicity and low toxicity have rendered the liposomal system highly attractive for drug delivery. This review focuses on the use of liposomes for the delivery of antioxidants in the prevention or treatment of several pathological conditions linked to oxidative stress.

Key words: delivery systems, free radicals, liposomes, oxidative stress

M.R. Mozafari (ed.), Nanocarrier Technologies: Frontiers of Nanotherapy, 191–205.

1.   INTRODUCTION

In the past few decades, there is increasing evidence connecting oxidative stress with a variety of pathological disorders including cancer, cardio-vascular diseases, chronic inflammatory disease, post-ischaemic organ injury, diabetes mellitus, and rheumatoid arthritis [1-4]. Also, a variety of anticancer drugs and xenobiotics are known to exert their therapeutic and/or toxic effects via the generation of reactive oxygen species [5-9]. Recently, it is becoming clear that reactive oxygen species are involved in the pathways, which convey both extracellular and intracellular signals to the nucleus, under a variety of pathophysiological conditions [3, 10, 11].

In general, oxidative stress has been characterized by an elevation in the steady-state concentration of reactive oxygen species including, but certainly not limited to, superoxide anion, hydrogen peroxide and hydroxyl radical [12, 13]. These reactive oxygen species are generated from the incomplete reduction of molecular oxygen. Enzymatic processes, such as the electron transport chain in the mitochondria, xanthine oxidase in ischemia-reperfusion, and cytochrome P-450-dependent activation of xenobiotics are known to play an important role in the generation of reactive oxygen species [3, 12]. Release of reactive oxygen species has been demonstrated in the respiratory burst of neutrophils and macrophages [10, 11].

The superoxide anion is formed by the univalent reduction of molecular oxygen. This process is mediated by enzymes such as xanthine oxidase and NAD(P)H oxidase or nonenzymically by redox-reactive compounds such as the semi-ubiquinone compound of the mitochondrial electron transport chain [10, 12, 13]. Superoxide dismutases convert the superoxide anion into hydrogen peroxide which in the presence of reduced transition metals (e.g. ferrous or cuprous ions) is converted to the highly reactive hydroxyl radical. In addition to the superoxide anion, hydrogen peroxide, and hydroxyl radical, there are other reactive oxygen species including nitric oxide, lipid radicals, peroxynitrite, and hypochlorous acid [10, 12, 13]. All these reactive oxygen species are potentially very reactive molecules and are known to exert their adverse effects by interacting with proteins, nucleic acids, carbohydrates, and lipids resulting in cell necrosis [3, 10-13].

Reactive oxygen species have also been implicated in triggering apoptosis or programmed cell death [3, 10, 14]. Apoptotic cell death is characterized by controlled autodigestion of the cell with the end result being the breakdown of the cell into membrane-bound fragments or apoptotic bodies, in most tissues being phagocytosed by adjacent cells. Activated neutrophils responding to inflammatory stimulation produce reactive oxygen species which attack neighboring cells triggering apoptosis.

Changes in the cellular redox potentials, depletion of reduced glutathione, and decreases in reducing equivalents, such as NADH and NADPH in the mitochondria by oxidants are also known to cause programmed cell death [15-17].

2.   ANTIOXIDANT DEFENSE SYSTEM

Reactive oxygen species exist in biological cells and tissues at low and measurable concentrations which are determined by the balance between their rates of production and their rates of clearance by various antioxidant compounds and antioxidant enzymes. Thus, any imbalances in the production or clearance of reactive oxygen species will result in organ or tissue injuries and these may occur when: increased generation of reactive oxygen species overwhelms the defence system; the defence system is severely compromised and incapable of detoxifying the normal flux or reactive metabolites; some combination of increased production and decreased detoxication occurs [12, 18, 19].

According to Gutteridge and Halliwell [19], antioxidants have been defined as substances that are able, at relatively low concentrations, to compete with other oxidizable substrates and, thus, significantly delay or inhibit the oxidation of these substrates. Substances that can protect against the damaging effects of reactive oxygen species can be included in the following categories: i) those aimed at preventing the generation and distribution of reactive oxygen species (the effective control of iron distribution and the destruction of peroxides by catalase or by glutathione peroxidase are included in this category); ii) those aimed at reactive metabolite scavenging including the maintenance of effective levels of antioxidants, such as vitamin E, vitamin C, β-carotene and glutathione, as well as the enzyme superoxide dismutase; and, iii) those aimed at free radical repair, particularly the maintenance of effective levels of glutathione [12, 18-20].

The major antioxidant enzymes found in living cells include the superoxide dismutase, catalase and glutathione peroxidase. Three different types of superoxide dismutase (SOD) have been isolated and characterized: a copper and zinc containing form (Cu-Zn-SOD) that is localized in the cytosol; a manganese-containing SOD (MnSOD) localized in the mitochondria; and, an extracellular form (EC-SOD) in the extracellular matrix. Catalase, which catalyzes the detoxication of hydrogen peroxide to water and oxygen, is localized mainly in peroxisomes, although some activity has been detected in mitochondria and cell cytoplasm. Glutathione peroxidase plays a major role in the detoxication of hydrogen peroxide, other hydroperoxides, and lipid peroxides via the glutathione redox cycle [12, 19].

The major non-enzymatic antioxidants include vitamin C (ascorbic acid), vitamin E (tocopherols and tocotrienols), beta-carotene and other micro-nutrients. Vitamin E, the principal antioxidant in the body, is composed of four tocopherols and four tocotrienols and, due to its lipophilicity, it is present in all cellular membranes [21]. Vitamin E neutralizes the highly reactive singlet oxygen molecules and protects polyunsaturated fatty acids in cell membranes from peroxidation. Ascorbic acid, a water soluble vitamin, is effective in scavenging free radicals, including hydroxyl radical, aqueous peroxyl radicals, and superoxide anion and is considered to be one of the most important antioxidants in extra cellular fluids. Ascorbic acid acts as a two-electron reducing agent and confers protection by contributing an electron to reduce free radicals, thus neutralizing these compounds in the extracellular aqueous environment prior to their reaction with biological molecules [18, 22, 23]. Glutathione (GSH) is the most abundant non-protein thiol in living organisms and plays a crucial role in intracellular protection against toxic compounds, such as reactive oxygen species and other free radicals [24, 25]. Glutathione can function as a nucleophile to form conjugates with many xenobiotic compounds and/or their metabolites and can also serve as a reductant in the metabolism of hydrogen peroxide and other organic hydroperoxides, a reaction catalyzed by glutathione peroxidases found in cytosols and mitochondria of various cells [24, 25].

3.   ANTIOXIDANT THERAPY

Antioxidant therapy has been defined as any treatment that prevents or decreases the adverse effects of reactive oxygen species. One of the strategies for pharmacological modification of oxidant-mediated tissue injury focuses on increasing the antioxidant capacity of cells or prevent the generation of reactive oxygen species. The ability of exogenous antioxidants to protect tissues from oxidant stress *in vivo* depends on the antioxidant used; its concentration at the site of action; the route of administration; and, the nature of the oxidant stress [26].

The benefit of antioxidants, such as superoxide dismutase, catalase, α-tocopherol, and glutathione, in the treatment of oxidant-induced tissue injury has been explored by several investigators. Most antioxidants, when applied directly and at relatively high concentrations to cellular systems *in vitro* , are effective in conferring protection against oxidant insults. On the other hand, studies in animals and humans have revealed that several antioxidants provide only modest benefit and have yet to be rendered into dependable and safe antioxidant therapies. The failure of antioxidants to seriously modify the injurious actions of oxidants has been attributed mostly to their inability to cross cell membrane barriers and/or to their rapid clearance from cells. Also, exposure of animals to the antioxidant enzymes (superoxide dismutase and catalase) has been encountered with antigenicity and immunogenecity problems. Experimental evidence, therefore, supports the necessity for the development of formulations that would enhance the delivery and retention of antioxidants in the tissues. In recent years, it has been demonstrated that the encapsulation of antioxidants in liposomes improves their therapeutic potential against oxidant-induced tissue injuries because liposomes presumably facilitate intracellular delivery and prolong the retention time of entrapped agents inside the cell [26-30].

4.   LIPOSOMES

In 1964, Bangham and Horne first demonstrated that the dispersion of naturally occurring phospholipids in an aqueous medium gives a vesicle (liposome) which has an interior water phase surrounded by lipid bilayer membrane. Since then, liposomes have been considered to be excellent models of cell membranes and have been employed as potent drug carriers in which various materials such as drugs, toxins, proteins, enzymes, antibodies and nucleotides are encapsulated [31-34]. The interest in liposomes as a drug delivery system stems from a combination of their relative innocuousness (composed of naturally-occurring, biodegradable compounds), their ability to diminish toxic effects of drugs, and their capability to increase the pharmacological activity of drugs which subsequently decrease the dosage of drugs [29, 35, 36].

Liposomes are artificially prepared phospholipid vesicles with amphipathic features. Hydrophilic molecules can be encapsulated in the aqueous spaces and lipophilic molecules can be incorporated in the lipid bilayers. Liposomes are considered as an acceptable and superior drug delivery system because they are biocompatible, biodegradeable and non-toxic. With respect to treating oxidant induced tissue injuries, it has been demonstrated that encapsulation of antioxidants in liposomes promotes their therapeutic potential against oxidant-induced tissue injuries, presumably by liposomes facilitating the intracellular uptake and extending the half-lives of the encapsulated antioxidants [27, 37-41].

4.1   Liposomes as a model membrane system for the testing of antioxidant agents

The membrane phospholipids contain variable amounts of polyun-saturated fatty acids, which are susceptible to oxidative damage. In biological membranes, this lipid peroxidation process leads to structural changes that alter membrane fluidity, permeability, and internal–external equilibrium, with consequent membrane destruction [42, 43]. Liposomes have been used extensively as an artificial membrane system to examine and compare the antioxidant properties of several agents [44-47]. The use of liposomal membranes for the examination of antioxidant properties of several compounds against an oxidant insult appears to be advantageous over the endogenous membrane systems (i.e. microsomes, mitochondria). Endogenous biological membranes contain antioxidants such as α-tocopherol that may influence the outcome of the experiment. For example, the presence of α-tocopherol in microsomal membranes enhances the antioxidant effect of certain antioxidants such as ascorbic acid or GSH, because of the ability of these antioxidants to regenerate the oxidized α-tocopherol [48]. Also, microsomes contain metabolizing enzymes that may modulate the antioxidant actions of compounds by either activating them or detoxify them.

The composition of the liposomal membrane systems prepared from phospholipids can be fully controlled and may be used to simulate the lipid composition of individual cellular membranes that often exhibit distinct lipid composition. As an example, the neuronal membranes are rich in sphingomyelin while the liver mitochondrial membranes contain phosphatidylcholine, lysophosphatidylcholine, diacylglycerol, and phosphatidylethanolamine [49-51].

4.2   Liposomes for the delivery of antioxidants

Several investigators are working towards the development of drug delivery systems that would result in the selective delivery of antioxidants and other therapeutic drugs to different tissues in sufficient concentrations to ameliorate oxidant-induced tissue injuries. The relative ease in incorporating hydrophilic and lipophilic therapeutic agents in liposomes; the properties of liposomes to protect therapeutic agents from inactivation; the possibility of directly delivering liposomes to an accessible body site such as the lung or liver; and, the relative non-immunogenicity and low toxicity of liposomes have rendered the liposomal system highly attractive for drug delivery [26, 27]. Liposomes can facilitate intracellular delivery of several therapeutic agents via fusion with the plasma membrane lipids, receptor-mediated endocytosis, and phagocytosis [29, 35, 36].

In a large number of studies, liposomes have been used extensively as a model of a lipid membrane system used for the evaluation of the antioxidant properties of several lipophilic and hydrophilic antioxidants against oxidant insults. In addition, but to a lesser extent, studies have also examined the use of liposomes for the delivery of water-soluble and lipid-soluble antioxidants as well as antioxidant enzymes to different organs and tissues for the treatment of oxidative stress-induced damage. Up-to-date, delivery of molecules with antioxidant properties to different organs and tissues include the lipophilic antioxidants α-tocopherol and CoQ10, the hydrophilic antioxidants glutathione, N-acetylcysteine, and quercetin, and the anti-oxidant enzymes superoxide dismutase and catalase [9, 37, 38, 52-54].

The use of the antioxidant enzymes superoxide dismutase and catalase for the treatment of oxidant-induced injuries has been discouraging. The lack of effectiveness by superoxide dismutase or catalase to protect against oxidative stress-induced damage has been attributed to their unfavourable pharmacokinetic profiles and physicochemical properties. In the case of superoxide dismutase, this enzyme cannot enter the target cell membrane because of its high molecular mass (which prevents intracellular transport) or its charge (which prevents its adherence to targets) [37, 38, 55]. However, the internalization of superoxide dismutase and/or catalase via liposomal delivery is shown to protect cells against the damaging actions of reactive oxygen species [37, 38, 56, 57]. Results from studies with animals indicate that administration of liposomally-encapsulated superoxide dismutase and catalase prolongs their circulating half-life [40, 55, 58]. Parenteral systemic administration of liposomal derivatives of superoxide dismutase provide enhanced protection in animal models of toxic hepatic necrosis [59], arthritis [40, 58], and cerebral ischemia/reperfusion injury [60-62]. Preliminary results from a limited number of clinical studies suggest that liposomal superoxide dismutase might protect against radiation-induced fibrosis [63, 64]. Subgingival application of liposome-encapsulated superoxide dismutase with scaling and root planing suppressed peridontal inflammation on experimentally induced periodontitis in beagle dogs [65]. Intratracheally administered surfactant liposomes, encapsulating CuZn-superoxide dis-mutase and catalase, increases the alveolar type II cell antioxidant activity and protects cells against oxidant stress in the lungs of premature or newborn animals [41, 66] or against bleomycin-induced lung injury [67].

The improved antioxidant effectiveness of liposomal-entrapped glutathione (GSH) over the free GSH has been demonstrated in paraquat- and hyperoxia-induced acute lung injuries. Smith *et al.* (1992) showed that intratracheal instillation of liposomal GSH faired better than free GSH in protecting against hyperoxia-induced lung injury. In another study, Suntres and Shek [68], showed that intratracheal instillation of liposome-entrapped GSH yielded a better protection than free GSH against paraquat-induced lung injury. The improved protection conferred by the liposomal GSH formulation was attributed to the extended retention of liposomes in the lung, thus allowing a slow release of its GSH content. Liposome encapsulation has shown to alter the pulmonary retention of GSH, with 18% and 10% of the dose administered remaining in the lung 24 and 48h post-treatment, whereas only 1-2% of the dose administered as free GSH was recovered in the lung 24h post-treatment [69, 70]. Similarly, intratracheal liposomal glutathione instillation in ventilated preterm infants raised the pulmonary glutathione and significantly reduced the levels of lipid peroxidation products [71]. Another thiol-containing compound, N-acetylcysteine (NAC), when delivered in its free form did not protect against the prolonged shock-induced acute lung injury while when administered as a liposomal formulation directly to the lungs of animals protected against the lung injury, a treatment effect occurring at low NAC doses [53]. Once again, the protective effect of NAC against the pulmonary damage following shock was attributed to the prolonged retention of liposomes in the lung, thus allowing a slow release of its NAC.

The role of α-tocopherol in modulating oxidant-induced cellular injury, permitting cells with high levels of the antioxidant to become more resistant to oxidative insults has been conclusive. It has been demonstrated that the administration of vitamin E, prior to an oxidative challenge, reduces the level of lipid peroxidation in several tissues and improves survival. However, in contrast to our studies, where the intratracheal instillation of α-tocopherol liposomes conferred a significant protective effect against the acute lung injuries induced by paraquat and other oxidants such as bleomycin, phorbol myristate acetate and lipopolysaccharide [26, 68, 72-80], the administration of non-liposomal α-tocopherol to animals by the oral or parenteral routes offered limited or no protection against paraquat or lipopolysaccharide-induced lung damage [81-85]. The apparent difference in the α-tocopherol effect between these studies may well be due to a difference in α-tocopherol concentration delivered to the lung. The intratracheal instillation of α-tocopherol liposomes achieved a substantially higher antioxidant level in the lung, approximately 1 mg/g lung weight [39] while the amount of antioxidant recovered from the lungs of animals after oral or parenteral administration of vitamin E was less than 40 μg/g lung tissue [86, 87]. Vitamin E, used as oral supplements, is often in the form of tocopheryl esters which are highly stable to oxidation but are absorbed only after they have been unesterified by the intestinal esterases. The active form of α-tocopherol is highly viscous oil, practically insoluble in water and readily oxidized by atmospheric oxygen. Attempts were made to deliver α-tocopherol to the lungs of animals by other means but results from these studies showed that organic solvents, such as ethanol and dimethyl-sulphoxide, used to solubilize the viscous free α-tocopherol, were toxic to the lung, and emulsifying agents and detergents, such as polyethylene glycol and Tween 80, caused respiratory failure, possibly by disrupting surface tension of the lung [39].

Data from several laboratrories have demonstrated that CoQ10 pretreatment protects the myocardium from ischemia-reperfusion injury via both antioxidant and bioenergetic pathways [52]. CoQ10 is an essential cofactor in the mitochondrial respiratory chain responsible for oxidative phosphorylation and can function as an antioxidant by acting as an oxygen free radical scavenger and thus stabilizing mitochondrial membranes and by inhibiting the arachidonic acid metabolic pathway and the formation of various prostaglandins [88]. However, its clinical use during an acute ischemic attack is hampered by the fact that following oral administration, it takes days to increase the CoQ10 blood and possibly tissue levels. The intravenous administration of CoQ10 as a liposomal formulation raised the serum and myocardial levels of CoQ10 and significantly improved the function and efficiency of myocardial tissue and reduced oxidant injury observed following ischemia-reperfusion [89].

Pretreatment of animals with the flavonoid quercetin was ineffective in protecting against carbon tetrachloride-induced hepatotoxicity. Carbon tetrachloride is known to mediate its liver toxicity via free-radical mechanisms including oxidative stress. Pretreatment of animals with mannosylated liposomal quercetin significantly lowered the hepatotoxic effect of carbon tetrachloride [54]. In another study, it was demonstrated that quercetin-filled liposomes improved the protective effects of the antioxidant against peroxynitrite-induced myocardial injury in isolated cardiac tissues and anesthetized animals [90].

It has been shown that the administration of liposomes containing more than one antioxidant is more advantageous in ameliorating oxidant-induced tissue injuries [55, 76, 91]. The antioxidant effect of liposomal formulations containing the enzymes SOD and catalase is more effective than those containing a single antioxidant enzyme [55, 57]. The therapeutic efficacy of an antioxidant liposome formulation, containing a lipophilic antioxidant, α-tocopherol, can be improved by encapsulating another antioxidant, such as GSH or L-ascorbic acid, in the same liposome preparation. These liposomes, containing both α-tocopherol and GSH were shown to be more effective in protecting against oxidant-induced lung injuries and lipid peroxidation than those containing either α-tocopherol or GSH alone presumably by α-tocopherol scavenging free radicals and stabilizing biological membranes while GSH, in addition to its ability to act as a free radical scavenger, can also regenerate α-tocopherol from its oxidized form [76]. Similarly, liposomes containing both α-tocopherol and L-ascorbic acid were more effective in preventing the generation of conjugated diene in cerebral tissues by the induction of global cerebral ischemia and reperfusion than those containing L-ascorbic acid through a synergistic action [91].

5.   FUTURE PERSPECTIVES

In summary, data from several studies clearly indicate that liposomes are highly efficient in terms of facilitating antioxidant delivery and achieving prophylactic and therapeutic efficacies against oxidative stress-induced damage. Because a large number of diseases are associated with the presence of reactive oxygen species, antioxidant liposomes have enormous health-related significance. Future studies should address the development of liposomes containing antioxidants and/or antimicrobials and antiinflam-matories, that preferentially accumulate at disease sites, such as those for infection and inflammation.

6.   REFERENCES

1.   Sies, H. (1985). Oxidative stress: Introductory remarks. In: Oxidative Stress. (Sies, H., Ed), pp. 1-8. Academic Press Inc., New York.
2.   Willcox, J. K., Ash, S. L. & Catignani, G. L. (2004). Antioxidants and prevention of chronic disease. Crit Rev Food Sci Nutr 44, 275-295.
3.   Djordjevic, V. B. (2004). Free radicals in cell biology. Int Rev Cytol 237, 57-89.
4.   Stohs, S. J. (1995). The role of free radicals in toxicity and disease. J Basic Clin Physiol Pharmacol 6, 205-228.
5.   Pelicano, H., Carney, D. & Huang, P. (2004). ROS stress in cancer cells and therapeutic implications. Drug Resist Updat 7, 97-110.
6.   Kovacic, P. & Osuna, J. A., Jr. (2000). Mechanisms of anti-cancer agents: emphasis on oxidative stress and electron transfer. Curr Pharm Des 6, 277-309.
7.   Doroshow, J. H. (1986). Role of hydrogen peroxide and hydroxyl radical formation in the killing of Ehrlich tumor cells by anticancer quinones. Proc Natl Acad Sci USA 83, 4514-4518.
8.   Stohs, S. J. & Bagchi, D. (1995). Oxidative mechanisms in the toxicity of metal ions. Free Radic Biol Med 18, 321-336.
9.   Suntres, Z. E. (2002). Role of antioxidants in paraquat toxicity. Toxicology 180, 65-77.
10. Forman, H. J. & Torres, M. (2002). Reactive oxygen species and cell signaling: respiratory burst in macrophage signaling. Am J Respir Crit Care Med 166, S4-8.
11. Di Virgilio, F. (2004). New pathways for reactive oxygen species generation in inflammation and potential novel pharmacological targets. Curr Pharm Des 10, 1647-1652.
12. Sies, H. (1987). Antioxidant activity in cells and organs. Am Rev Respir Dis 136, 478-480.
13. Toyokuni, S. (1999). Reactive oxygen species-induced molecular damage and its application in pathology. Pathol Int 49, 91-102.

14. Kadenbach, B., Arnold, S., Lee, I. *et al.* (2004). The possible role of cytochrome c oxidase in stress-induced apoptosis and degenerative diseases. Biochim Biophys Acta 1655, 400-408.

15. Zoratti, M. & Szabo, I. (1995). The mitochondrial permeability transition. Biochim Biophys Acta 1241, 139-176.

16. Bernardi, P. (1996). The permeability transition pore. Control points of a cyclosporin A-sensitive mitochondrial channel involved in cell death. Biochim Biophys Acta 1275, 5-9.

17. Bernardi, P. & Petronilli, V. (1996). The permeability transition pore as a mitochondrial calcium release channel: a critical appraisal. J Bioenerg Biomembr 28, 131-138.

18. Evans, P. & Halliwell, B. (2001). Micronutrients: oxidant/antioxidant status. Br J Nutr 85 Suppl 2, S67-74.

19. Gutteridge, J. M. & Halliwell, B. (2000). Free radicals and antioxidants in the year 2000. A historical look to the future. Ann N Y Acad Sci 899, 136-147.

20. Suntres, Z. E., Omri, A. & Shek, P. N. (2002). Pseudomonas aeruginosa-induced lung injury: role of oxidative stress. Microb Pathog 32, 27-34.

21. Chow, C. K. (2004). Biological functions and metabolic fate of vitamin E revisited. J Biomed Sci 11, 295-302.

22. Halliwell, B. (1996). Vitamin C: antioxidant or pro-oxidant *in vivo*? Free Radic Res 25, 439-454.

23. Carr, A. & Frei, B. (1999). Does vitamin C act as a pro-oxidant under physiological conditions? Faseb J 13, 1007-1024.

24. Anderson, M. E. (1997). Glutathione and glutathione delivery compounds. Adv Pharmacol 38, 65-78.

25. Anderson, M. E. & Luo, J. L. (1998). Glutathione therapy: from prodrugs to genes. Semin Liver Dis 18, 415-424.

26. Suntres, Z. E. & Shek, P. N. (1995). Liposome-associated antioxidants for pulmonary applications. In Liposomes in Biomedical Applications. (Shek, P., Ed), pp. 179-198. Harwood Academic Publishers, Singapore.

27. Stone, W. L. & Smith, M. (2004). Therapeutic uses of antioxidant liposomes. Mol Biotechnol 27, 217-230.

28. Shek, P. N., Suntres, Z. E. & Brooks, J. I. (1994). Liposomes in pulmonary applications: physicochemical considerations, pulmonary distribution and antioxidant delivery. J Drug Target 2, 431-442.

29. Allen, T. M. (1998). Liposomal drug formulations. Rationale for development and what we can expect for the future. Drugs 56, 747-756.

30. Langner, M. & Kral, T. E. (1999). Liposome-based drug delivery systems. Pol J Pharmacol 51, 211-222.

31. Gilbert, B. E. (1996). Liposomal aerosols in the management of pulmonary infections. J Aerosol Med 9, 111-122.

32. Taylor, M. G. & Farr, S. J. (1993). Liposomes for drug delivery to the respiratory tract. Drug Develop. Indust. Pharm. 19, 123-142.

33. Kimelberg, H. K. & Mayhew, E. G. (1978). Properties and biological effects of liposomes and their uses in pharmacology and toxicology. CRC Crit. Rev. Toxicol. 96, 25-79.

34. Yatvin, M. B. & Lelkes, P. I. (1982). Clinical prospects for liposomes. Med Phys 9, 149-175.

35. Gregoriadis, G. (1991). Overview of liposomes. J Antimicrob Chemother 28 Suppl B, 39-48.

36. Gregoriadis, G. (1995). Engineering liposomes for drug delivery: progress and problems. Trends Biotechnol 13, 527-537.

37. Muzykantov, V. R. (2001). Targeting of superoxide dismutase and catalase to vascular endothelium. J Control Release 71, 1-21.

38. Muzykantov, V. R. (2001). Delivery of antioxidant enzyme proteins to the lung. Antioxid Redox Signal 3, 39-62.

39. Suntres, Z. E., Hepworth, S. R. & Shek, P. N. (1993). Pulmonary uptake of liposome-associated alpha-tocopherol following intratracheal instillation in rats. J Pharm Pharmacol 45, 514-520.

40. Corvo, M. L., Boerman, O. C., Oyen, W. J. *et al.* (1999). Intravenous administration of superoxide dismutase entrapped in long circulating liposomes. II. *In vivo* fate in a rat model of adjuvant arthritis. Biochim Biophys Acta 1419, 325-334.

41. Walther, F. J., David-Cu, R. & Lopez, S. L. (1995). Antioxidant-surfactant liposomes mitigate hyperoxic lung injury in premature rabbits. Am J Physiol 269, L613-617.

42. Watanabe, H., Kobayashi, A., Yamamoto, T. *et al.* (1990). Alterations of human erythrocyte membrane fluidity by oxygen-derived free radicals and calcium. Free Radic Biol Med 8, 507-514.

43. Cimato, A. N., Piehl, L. L., Facorro, G. B. *et al.* (2004). Antioxidant effects of water- and lipid-soluble nitroxide radicals in liposomes. Free Radic Biol Med 37, 2042-2051.

44. Sengupta, B., Banerjee, A. & Sengupta, P. K. (2004). Investigations on the binding and antioxidant properties of the plant flavonoid fisetin in model biomembranes. FEBS Lett 570, 77-81.

45. Zhang, Y., Cichewicz, R. H. & Nair, M. G. (2004). Lipid peroxidation inhibitory compounds from daylily (Hemerocallis fulva) leaves. Life Sci 75, 753-763.

46. Zago, M. P. & Oteiza, P. I. (2001). The antioxidant properties of zinc: interactions with iron and antioxidants. Free Radic Biol Med 31, 266-274.

47. Junghans, A., Sies, H. & Stahl, W. (2001). Carotenoid-Containing Unilamellar Liposomes Loaded with Glutathione: a Model to Study Hydrophobic-Hydrophilic Antioxidant Interaction. Free Radic Res 33, 801-808.

48. Leedle, R. A. & Aust, S. D. (1990). The effect of glutathione on the vitamin E requirement for inhibition of liver microsomal lipid peroxidation. Lipids 25, 241-245.

49. Yeagle, P. L. (1993). Phosphorus-31 nuclear magnetic resonance in membrane fusion studies. Methods Enzymol 220, 68-79.

50. Yegin, A., Akbas, S. H., Ozben, T. *et al.* (2002). Secretory phospholipase A2 and phospholipids in neural membranes in an experimental epilepsy model. Acta Neurol Scand 106, 258-262.

51. de Lima, V. R., Morfim, M. P., Teixeira, A. *et al.* (2004). Relationship between the action of reactive oxygen and nitrogen species on bilayer membranes and antioxidants. Chem Phys Lipids 132, 197-208.

52. Crestanello, J. A., Kamelgard, J., Lingle, D. M. *et al.* (1996). Elucidation of a tripartite mechanism underlying the improvement in cardiac tolerance to ischemia by coenzyme Q10 pretreatment. J Thorac Cardiovasc Surg 111, 443-450.

53. Fan, J., Shek, P. N., Suntres, Z. E. *et al.* (2000). Liposomal antioxidants provide prolonged protection against acute respiratory distress syndrome. Surgery 128, 332-338.

54. Mandal, A. K., Sinha, J., Mandal, S. *et al.* (2002). Targeting of liposomal flavonoid to liver in combating hepatocellular oxidative damage. Drug Deliv 9, 181-185.

55. Freeman, B. A., Turrens, J. F., Mirza, Z. *et al.* (1985). Modulation of oxidant lung injury by using liposome-entrapped superoxide dismutase and catalase. Fed Proc 44, 2591-2595.

56. Beckman, J. S., Minor, R. L., Jr. & Freeman, B. A. (1986). Augmentation of antioxidant enzymes in vascular endothelium. J Free Radic Biol Med 2, 359-365.

57. Turrens, J. F., Crapo, J. D. & Freeman, B. A. (1984). Protection against oxygen toxicity by intravenous injection of liposome-entrapped catalase and superoxide dismutase. J Clin Invest 73, 87-95.

58. Cruz, M. E., Manuela Gaspar, M., Barbara, M. *et al.* (2005). Liposomal superoxide dismutases and their use in the treatment of experimental arthritis. Methods Enzymol 391, 395-413.

59. Nakae, D., Yamamoto, K., Yoshiji, H. *et al.* (1990). Liposome-encapsulated superoxide dismutase prevents liver necrosis induced by acetaminophen. Am J Pathol 136, 787-795.

60. Homi, H. M., Freitas, J. J., Curi, R. *et al.* (2002). Changes in superoxide dismutase and catalase activities of rat brain regions during early global transient ischemia/reperfusion. Neurosci Lett 333, 37-40.

61. Chan, P. H. (1992). Antioxidant-dependent amelioration of brain injury: role of CuZn-superoxide dismutase. J Neurotrauma 9 Suppl 2, S417-423.

62. Stanimirovic, D. B., Markovic, M., Micic, D. V. *et al.* (1994). Liposome-entrapped superoxide dismutase reduces ischemia/reperfusion 'oxidative stress' in gerbil brain. Neurochem Res 19, 1473-1478.

63. Baillet, F., Housset, M., Michelson, A. M. *et al.* (1986). Treatment of radiofibrosis with liposomal superoxide dismutase. Preliminary results of 50 cases. Free Radic Res Commun 1, 387-394.

64. Delanian, S., Baillet, F., Huart, J. *et al.* (1994). Successful treatment of radiation-induced fibrosis using liposomal Cu/Zn superoxide dismutase: clinical trial. Radiother Oncol 32, 12-20.

65. Petelin, M., Pavlica, Z., Ivanusa, T. *et al.* (2000). Local delivery of liposome-encapsulated superoxide dismutase and catalase suppress periodontal inflammation in beagles. J Clin Periodontol 27, 918-925.

66. Tanswell, A. K. & Freeman, B. A. (1987). Liposome-entrapped antioxidant enzymes prevent lethal O2 toxicity in the newborn rat. J Appl Physiol 63, 347-352.

67. Ledwozyw, A. (1991). Protective effect of liposome-entrapped superoxide dismutase and catalase on bleomycin-induced lung injury in rats. I. Antioxidant enzyme activities and lipid peroxidation. Acta Vet Hung 39, 215-224.

68. Suntres, Z. E. & Shek, P. N. (1995). Liposomal alpha-tocopherol alleviates the progression of paraquat-induced lung damage. J Drug Target 2, 493-500.

69. Suntres, Z. E. & Shek, P. N. (1994). Incorporation of alpha-tocopherol in liposomes promotes the retention of liposome-encapsulated glutathione in the rat lung. J Pharm Pharmacol 46, 23-28.

70. Smith, L. J., Anderson, J. & Shamsuddin, M. (1992). Glutathione localization and distribution after intratracheal instillation. Implications for treatment. Am Rev Respir Dis 145, 153-159.

71. Cooke, R. W. & Drury, J. A. (2005). Reduction of oxidative stress marker in lung fluid of preterm infants after administration of intra-tracheal liposomal glutathione. Biol Neonate 87, 178-180.

72. Suntres, Z. E., Hepworth, S. R. & Shek, P. N. (1992). Protective effect of liposome-associated alpha-tocopherol against paraquat-induced acute lung toxicity. Biochem Pharmacol 44, 1811-1818.

73. Suntres, Z. E. & Shek, P. N. (1995). Intratracheally administered liposomal alpha-tocopherol protects the lung against long-term toxic effects of paraquat. Biomed Environ Sci 8, 289-300.

74. Suntres, Z. E. & Shek, P. N. (1995). Prevention of phorbol myristate acetate-induced acute lung injury by alpha-tocopherol liposomes. J Drug Target 3, 201-208.

75. Suntres, Z. E. & Shek, P. N. (1996). Treatment of LPS-induced tissue injury: role of liposomal antioxidants. Shock 6 Suppl 1, S57-64.

76. Suntres, Z. E. & Shek, P. N. (1996). Alleviation of paraquat-induced lung injury by pretreatment with bifunctional liposomes containing alpha-tocopherol and glutathione. Biochem Pharmacol 52, 1515-1520.

77. Suntres, Z. E. & Shek, P. N. (1996). The pulmonary uptake of intravenously administered liposomal alpha-tocopherol is augmented in acute lung injury. J Drug Target 4, 151-159.

78. Suntres, Z. E. & Shek, P. N. (1997). Protective effect of liposomal alpha-tocopherol against bleomycin-induced lung injury. Biomed Environ Sci 10, 47-59.

79. Suntres, Z. E. & Shek, P. N. (1998). Prophylaxis against lipopolysaccharide-induced acute lung injury by alpha-tocopherol liposomes. Crit Care Med 26, 723-729.

80. Suntres, Z. E. & Shek, P. N. (1998). Liposomes promote pulmonary glucocorticoid delivery. J Drug Target 6, 175-182.

81. Ramazzotto, L. J. & Engstrom, R. (1975). Dietary vitamin E and the effects of inhaled nitrogen dioxide on rat lungs. Environ Physiol Biochem 5, 226-234.
82. Redetzki, H. M., Wood, C. D. & Grafton, W. D. (1980). Vitamin E and paraquat poisoning. Vet Hum Toxicol 22, 395-397.
83. Stephens, R. J., Buntman, D. J., Negi, D. S. *et al.* (1983). Tissue levels of vitamin E in the lung and the cellular response to injury resulting from oxidant gas exposure. Chest 83, 37S-39S.
84. Warren, D. L., Hyde, D. M. & Last, J. A. (1988). Synergistic interaction of ozone and respirable aerosols on rat lungs. IV. Protection by quenchers of reactive oxygen species. Toxicology 53, 113-133.
85. Yasaka, T., Okudaira, K., Fujito, H. *et al.* (1986). Further studies of lipid peroxidation in human paraquat poisoning. Arch Intern Med 146, 681-685.
86. Gallo-Torres, H. E. (1980). Transport and metabolism. Marcel Dekker, New York.
87. Knight, M. E. & Roberts, R. J. (1985). Tissue vitamin E levels in newborn rabbits after pharmacologic dosing: influence of dose, dosage form and route of administration. Dev. Pharmacol. Ther. 8, 96-106.
88. Crane, F. L. (2001). Biochemical functions of coenzyme Q10. J Am Coll Nutr 20, 591-598.
89. Niibori, K., Yokoyama, H., Crestanello, J. A. *et al.* (1998). Acute administration of liposomal coenzyme Q10 increases myocardial tissue levels and improves tolerance to ischemia reperfusion injury. J Surg Res 79, 141-145.
90. Soloviev, A., Stefanov, A., Parshikov, A. *et al.* (2002). Arrhythmogenic peroxynitrite-induced alterations in mammalian heart contractility and its prevention with quercetin-filled liposomes. Cardiovasc Toxicol 2, 129-139.
91. Sinha, J., Das, N. & Basu, M. K. (2001). Liposomal antioxidants in combating ischemia-reperfusion injury in rat brain. Biomed Pharmacother 55, 264-271.

Chapter 12

INTERACTION OF DENDRIMERS WITH MODEL LIPID MEMBRANES ASSESSED BY DSC AND RAMAN SPECTROSCOPY

Konstantinos Gardikis [1], Sophia Hatziantoniou [1], Kyriakos Viras [2], Matthias Wagner [3] and Costas Demetzos [1]

[1]Department of Pharmaceutical Technology, School of Pharmacy, Panepistimiopolis, Zografou 15771, University of Athens, Athens, Greece; [2]Department of Chemistry, Laboratory of Physical Chemistry, University of Athens, Athens, Greece; [3]Mettler-Toledo GmbH, Business Unit Analytical, Sonnenbergstrasse 74, CH-8603, Schwerzenbach, Switzerland

Abstract: The aim of the present work was to study the interaction between PAMAM generation 4 (G4) dendrimer with model lipid membranes (DPPC) for designing new controlled released systems for bioactive molecules by combining dendrimer and liposomal technologies. Thermal analysis and Raman spectroscopy were applied to assess the thermodynamic changes caused by PAMAM G4 (polyamidoamines) dendrimer and to specify the exact location of this dendrimer into the DPPC lipid bilayer. DSC thermograms indicated that the maximum percent of PAMAM G4 that can be incorporated in the DPPC membrane without deranging its integrity is 5%. The Raman intensity ratios $I_{2935/2880}$ and $I_{1090/1130}$ cm^{-1} showed the degree of the fluidity of the lipid bilayer, while the absorption at 715 cm^{-1} showed a strong interaction of PAMAM G4 with the polar head group of phospholipid. The results showed that the incorporation of the PAMAM G4 dendrimer in DPPC bilayers causes a concentration dependent increase of the membrane fluidity and they interact strongly with both the lipophilic part and the polar head group of the phospholipids. Additionally, due to the current weak knowledge of how dendrimers interact with lipidic membranes these results may justify the tendency of dendrimers to disrupt biological membranes.

Key words: Dendrimer, PAMAM, DSC, RAMAN spectroscopy.

M.R. Mozafari (ed.), Nanocarrier Technologies: Frontiers of Nanotherapy, 207–220.
© 2006 Springer. Printed in the Netherlands.

1.   INTRODUCTION

The major challenge in drug delivery is to control the distribution of bioactive molecules in various organs after systemic administration. The inefficient drug delivery to the target tissue and the side effects after administration is due to the lack of sufficient drug carriers that can influence not only pharmacokinetics but also biodistribution of the administered molecules. A large number of drugs have not been effective due to their inability to reach to the target tissue. Diseases like cancer, hormone deficiencies, microbial infections, etc. need effective drug carriers comprising such properties, which can retain their physicochemical characteristics within the biological media. It is obvious that the design and development of drug carries is a difficult issue because they have to behave physicochemically as biocolloidal systems after administration. Their physicochemical parameters should be under control and *in vitro* experiments have to be designed rationally in order to achieve such delivery systems that can provide essential breakthroughs in the fight against diseases.

Nanotechnology, (from the Greek word *νανο* : nano and *τεχνολογία* : technology) seems to be the current promising multidisciplinary scientific field, which covers issues from biology (cellular functions, membrane fusion), colloid science (interface phenomena, colloidal stability) chemistry (organic synthesis, catalysis), biochemistry and molecular biology (gene expression), physics, mathematics and medicine. Albert Frank characterized nanotechnology as *'the area of science and technology where dimensions and tolerances are in range of 0.1 to 100nm'* [1].

Numerous drug carriers such as micelles, mixed micelles, emulsions and nanoparticles are expected to solve the problem of drug delivery or drug instability in biological fluids or toxicity in the case of anticancer drugs. One of the most promising colloidal systems for drug delivery is liposomes which have been characterized as synthetic analogues of natural membranes [2]. Their utility as drug carriers first suggested by Weismann in 1969 [1] was based on their biocompatibility, colloidal character and incorporating properties. Liposomes consist mainly of phospholipids and form lipid bilayers, while their lipid composition, size distribution and their stability as well as number of lamellae and their morphological characteristics are essential parameters for their effectiveness as drug carriers [3-5]. Their physicochemical properties are the main determinants of their efficiency as they can improve the pharmacokinetic properties of bioactive molecules, which influence their pharmacological efficiency. Therefore, the stability of the membrane bilayers incorporating bioactive compounds and their physicochemical properties are important when designing liposomes as drug carriers.

Liposomes can be characterized as nanovectors, which are a class of nanotechnological devices for drug delivery, and they are the first class of nanovectors under development for more effective drug delivery. It is well known that polymers can act as nanovectors and a huge number have been investigated [6].

Dendrimers, are considered as highly branched macromolecules; they are small in size, while their low polydispersity can contribute to the reproducibility of their pharmacokinetic behavior [7, 8]. An ideal dendrimer as drug delivery system must be non-toxic, non-immunogenic and biodegradable [8]. The first complete dendrimer family which has been synthesized, characterized and commercialized is the Poly(amidoamine) (PAMAM) dendrimers. They are characterized as safe and non-immunogenic and they are used in drug delivery, delivery of antisense nucleotides and gene therapy, both *in vitro* and *in vivo* [9]. The bibliography information on the use of dendrimers as drug delivery carriers is limited and only few studies have been published concerning the interaction of dendrimers and drugs [10]. The use of dendrimers as modulators of the release of a drug incorporated into liposomes and the possible alterations of the drug bioavailability seems to be an attractive field for research.

The word dendrimer is derived from the Greek words dendri- (tree branch-like) and meros (part of), and the synthesis of '*true dendrimers*' was achieved by Tomalia group (Dow Chemical Co.) about 20 years ago [11, 12].

PAMAM G4 dendrimers constitute the first dendrimer family to be commercialized, and represent the most extensively characterized and best-understood series at this time. Dendrimers may be viewed as unique, information-processing, nanoscale devices. Each architectural component manifests a specific function, while at the same time defining properties for these nanostructures as they are grown generation by generation. A typical dendrimer consists of a core, the branches and the surface groups. The core may be thought of as the molecular information center from which size, shape, directionality, and multiplicity are expressed via the covalent connectivity to the outer shells. The branch-cell multiplicity determines the density and degree of amplification as an exponential function of generation. The interior composition and volume of solvent-filled void space determines the extent and nature of guest-host (endo-receptor) properties that are possible within a particular dendrimer family and generation. Finally the terminal groups may perform several functions, such as bonding of a chemical group or a reagent [13].

Dendrimers - due to their small size, low polydispersity and complete control of their structure - provide some special advantages for drug delivery. These include the prolongation of the drug in the circulation, the protection of the drug from its environment, the increase of the stability and, possibly, effectiveness of the drug and the capability of targeting to specific tissues. Their applications include their use in the chemical industry as catalysts, lubricants, battery components and slow reagent factors in perfumes but their most important applications are DNA delivery (gene delivery into cells) as well as delivery, targeting and release of other drugs [14].

In this work we studied the effect of PAMAM G4 dendrimer on lipid bilayers composed of DPPC in order to contribute to the knowledge in drawing new drug delivery systems that consist of dendrimers incorporating bioactive molecules attached to liposomes. The interaction between DPPC lipid bilayers and PAMAM G4 dendrimer, was assessed using Thermal Analysis (more specifically Differential Scanning Calorimetry) and Raman Spectroscopy.

2.   PHYSICAL METHODS FOR STUDYING THE CHANGES IN PHYSICAL PROPERTIES OF DPPC LIPID BILAYERS INCORPORATING PAMAM G4 DENDRIMER

2.1   DPPC as a building block of model lipid bilayers

The most common lipids that can be found in the cell membrane and that can be used for the development of lipoid drug delivery systems are categorized as *glycerolipids, sphingolipids* and *sterols*. In this study dipalmitoyl-phosphatidylcholine (DPPC, a subcategory of glycerolipids) has been used as an ingredient of model lipid membrane. This phospholipid consists of a molecule of glycerol which is esterified with two molecules of palmitic acid $[CH_3(CH_2)_{14}COOH]$ and with a phosphoric acid, which in turn is esterified with an amine (choline). As a consequence, DPPC consists of a lipophilic (palmitic acid) and a hydrophilic (choline) part. This amphiphilic property of DPPC enables it to take a very special orientation when dissolved in water - i.e. to form lipid bilayers with the polar head in the exterior creating hydrogen bonds with the water molecules and the lipid part in the interior.

Phospholipids exist in various forms, which are called mesomorphs. DPPC in aqueous environment exists in two totally different mesomorphic phases known as L_α and $L_{\beta'}$. The $L_{\beta'}$ crystalline form of DPPC corresponds to the *gel phase* while the L_α crystalline form corresponds to the *liquid crystalline state*. The transition from the gel phase to the liquid-crystalline phase can be achieved by temperature rising. In that case the DPPC molecules move more rapidly increasing the lateral and rotational diffusion among them, while the intermolecular motion around C-C bonds increases and, thus, kink forms are created. The *all-trans* conformation of DPPC in the gel phase can be leaded through kink conformations to a *gauche* form (120 degree rotation of the simple C-C bond) and formation of a *gauche-trans-gauche* conformation of the lipid chain.

Phosphatidylcholines with the choline polar head (e.g. DPPC) demonstrate an endothermic transition that happens prior to the main gel to liquid-crystalline transition. This thermotropic phenomenon is called *pre-transition* and creates bilayers with the lipid chains fully stretched and tilted ($L_{\beta'}$ phase). The tilt angle of the lipid chain is temperature dependable and is minimized at the pre-transition temperature. Then the crystalline phase is called ripple phase ($P_{\beta'}$) (Fig. 12-1). The exact knowledge and control of these thermodynamic properties of the lipid bilayers are of high importance in the development of control release systems because the encapsulation, stability and release of biologically active molecules depend totally on them [3, 15].

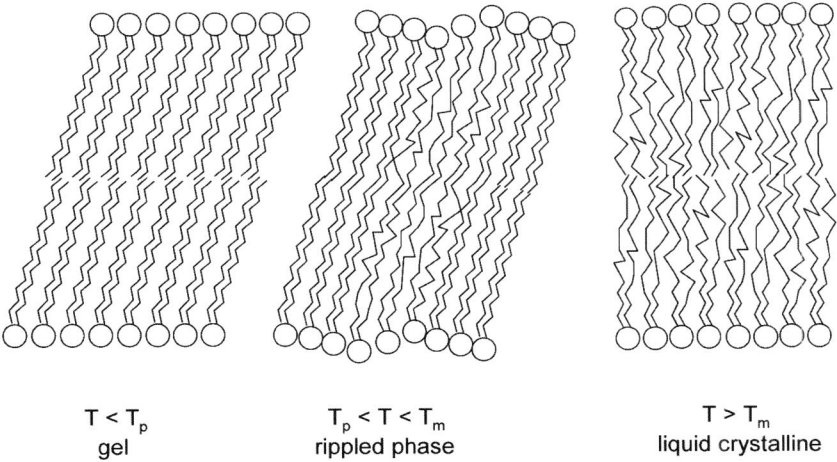

T < T$_p$	T$_p$ < T < T$_m$	T > T$_m$
gel	rippled phase	liquid crystalline

Figure 12-1. Schematic presentation of the alignment of acyl chains in a saturated phospholipid such as DPPC to form a double layer of molecules: gel in quasi-crystalline state ($T<T_p$), rippled phase ($T_p<T<T_m$) and in the liquid crystalline state ($T>T_m$). (Hatziantoniou, S., Demetzos, C. and Wagner, M. in Mettler-Toledo TA UserCom, Vol. 21, 1/2005).

2.2   Differential Scanning Calorimetry

A great number of physical and chemical changes can be produced by temperature changes and the methods that characterize these changes during the heating or the cooling of a specimen are referred to as *thermal analysis.* The most common thermal analysis methods are differential thermal analysis (DTA), thermogravimetric analysis (TGA), hot-stage microscopy (HSM), thermomechanical analysis (TMA), dynamic mechanical analysis (DMA) and differential scanning calorimetry (DSC) [16].

DSC measures the heat flow going into or released by a specimen. From that the heat capacity at constant pressure (C_p) can be calculated. Heat capacity units are cal $°C^{-1}$ or J $°C^{-1}$. It measures the amount of heat input (q) required to raise the temperature of a specimen by one degree Celsius while at constant pressure. Heat capacity is usually normalized by dividing the specimen heat capacity by the number of grams to get the heat required to rise one gram of specimen by one degree Celsius. This corresponds then to the specific heat capacity c_p. If desired, the number of moles can normalize heat capacity. When observing thermal transitions, we note that heat capacity is related to the second derivative of free energy:

$$(\partial G / \partial T)_p = - S - q / T$$

$$(\partial^2 G / \partial T^2) = - \{\partial(q/T) / \partial T)_p = (S / T) - (1 / T)(\partial q / \partial T)_p = (S / T) - C_p / T$$

and constant-pressure heat capacity is defined by:

$$C_p = (\partial q / \partial T)_p = S - T(\partial^2 G / \partial T^2)_p$$

(S is the entropy of the system and G is the free energy of the system and T is the temperature and q is the heat input).

The method is to carry out the reaction adiabatically (i.e. with no heat flow) and measure the change in temperature within the calorimeter. If the temperature changes from T_0 to T_1 the enthalpy of the reaction (ΔH) is:

$$\Delta H = {}^{T_1}\!\!\int_{T_0} CpdT$$

Usually ΔT is small and C_p is independent of temperature between T_0 and T_1. The integral thus reduces to:

$$\Delta H = C_p (T_1 - T_0) = C_p \Delta T$$

DSC has several characteristic temperatures, which may be used to describe it. The *onset temperature* (T_{onset}) is the temperature which is constructed by the intersection of the baseline tangent with the tangent of the leading edge of the endotherm or exotherm of a transition. The *peak temperature* (T_m) is the temperature represented by the apex of the peak. Where the peak transition returns to the baseline may refer to as the *recovery or endset temperature* (T_{endset}). $T_{1/2}$ is the temperature that corresponds to the half of the enthalpy change during the transition [16].

The phenomena that take place during heating process are as follows: The *glass transition* or T_g is a transition which occurs in amorphous or semicrystalline materials. Around T_g, C_p undergoes a quasi-discontinuous change from a lower value to a higher value (it is not sharp but happens over a broader range in temperature depending on the molecular structure of the material). If an amorphous material tends to crystallize or a semicrystalline material did not crystallize to the limit of its ability when cooled, it is possible to see crystallization during the heating. *Crystallization* shows up as an exothermic peak and the crystallization temperature T_c is always between T_g and the melting temperature T_m. Crystallization will only occur in samples that can crystallize and in samples that are not already crystallized as far as they can be crystallized. The highest temperature peak in a DSC thermogram is often the melting transition. *Melting* is a first order endothermic peak, which means that it requires heat. For small molecules the peak is very sharp while for larger molecules, such as polymers or lipid bilayers, the melting transition is broad. The area under a melting transition curve is the total amount of heat absorbed during the melting process. Further peaks can be attributed to loss of solvent molecules (evaporation, mostly endothermic) and chemical reactions (endo- or exothermic), among them decomposition.

The changes of the enthalpy as a function of phase transitions when an additive like dendrimer is incorporated into lipid bilayers have never been studied. The modifications on the ordered lipidic structure of DPPC bilayers probably due to the structural conformation of the additive, which can strongly interact either with the lipophilic part of DPPC and thus resulting greater thermotropic changes in T_m, or with the polar group of DPPC resulting an abolition of pre-transition peak at 36,14°C.

In this study DSC was applied in order to assess the thermodynamic changes in the lipid bilayers caused by the incorporated additives such as dendrimer (i.e. PAMAM G4) and thus further understand dendrimers' interaction with DPPC lipid bilayers and can be helpful designing drug carriers combined dendrimer and liposomal technologies [17].

2.3   Raman spectroscopy

When electromagnetic radiation of energy content hv irradiates a molecule, the energy may be transmitted, absorbed, or scattered. In the Tyndall effect the radiation is scattered by particles (smoke or fog, for example). In Rayleigh scattering the molecules scatter the light. No change in wavelength of individual photons occurs in either Tyndall or Rayleigh scattering. Raman spectroscopy is based on the Raman effect, which is the inelastic scattering of photons by molecules. Raman spectroscopy is actually the measurement of the wavelength and intensity of inelastically scattered light from molecules. The Raman scattered light occurs at wavelengths that are shifted from the incident light by the energies of molecular vibrations. The mechanism of Raman scattering is different from that of infrared absorption, and Raman and IR spectra provide complementary information. For a transition to be Raman active, there must be a change in polarizability of the molecule; the polarizability must change with the vibrational motion for that vibration to inelastically scatter radiation. The polarizability depends on how tightly the electrons are bound to the nuclei. In the symmetric stretch the strength of electron binding is different between the minimum and maximum internuclear distances. Therefore the polarizability changes during the vibration and this vibrational mode scatters Raman light. In the asymmetric stretch the electrons are more easily polarized in the bond that expands but are less easily polarized in the bond that compresses. There is no overall change in polarizability and the symmetric stretch is Raman inactive [18].

Typical applications of Raman spectroscopy are in quantitative analysis, structure determination and, most importantly in our study, multicomponent qualitative analysis. In this study Raman spectra helped us to specify the exact location of the PAMAM G4 dendrimer into the lipid bilayers and to clarify the mechanism of their interaction.

3.   MATERIALS AND METHODS

PAMAM G4 dendrimer was purchased from Aldrich. The dendrimer came as 10% w/w solutions in methyl alcohol and it was dried under reduced pressure. Then it was dissolved in pure methyl alcohol in order to obtain 10% w/v solutions. DPPC was purchased from Avanti Polar Lipids. Mixtures of DPPC and PAMAM G4 dendrimer, were prepared at 3%, 5%, 10%, 15%, 20% and 30% molar ratio. The mixture was dissolved in pure chloroform and dried under vacuum. HPLC grade deionized water filtered through Millipore Filters (pore size 200nm) was added (1:2 w/w) in the dried mixtures. A quantity of 4 to 7mg (total weight) was used for the DSC measurements while 10 to 12mg (total weight) was used for the Raman Spectroscopy.

Samples were prepared in 40 µl crucibles hermetically sealed with lid. Prior to DSC scanning, the samples underwent a quick heating and a quick cooling cycle (scanning rate of 10°C/min and 20°C/min, respectively) to ensure equilibration and exemption of the thermal history of the specimens. All samples were scanned three to four times until identical thermograms were obtained using a scanning rate of 2°C/min. As a reference sample an empty pan was used and the temperature scale of the calorimeter was calibrated using indium (T_m = 156,6°C). A Mettler Toledo DSC822^e heat-flux DSC was used.

Prior to Raman spectroscopy the hydrated mixture was left in the spectrometric cell for 16 to 24 hours to ensure equilibration. The temperature was 25°C. High-frequency Raman spectra were recorded with a Perkin-Elmer GX Fourier transform spectrometer. A diode pumped Nd:YAG laser at 1064nm was used as excitation source. The spectra were obtained at 4cm^{-1} resolution from 3500 to 500cm^{-1} with interval 2cm^{-1}. The laser power was controlled to be constant at 400mW during the experiments. Analysis of the spectra was carried out using GRAMS/32 data analysis software.

4. RESULTS AND DISCUSSION

Thermal analysis results are based on T_{onset}, T_m, $T_{1/2}$ and ΔH. Fully hydrated DPPC bilayers incorporating PAMAM G4 dendrimer showed thermograms consisting of broad enthalpy transitions, abolition of the pre-transition and reduction of the enthalpy change of the gel to liquid-crystalline phase transition of DPPC bilayers as shown in Table 12-1 and Fig. 12-2. At the concentration of 10% PAMAM G4 dendrimer, the homogeneity of the transition is lost, a fact that can be attributed to the apparition of 'domains' of PAMAM G4 high concentrations in the lipid bilayer. The abolition of the pre-transition peak at all concentrations can be attributed to interfacial interactions between the PAMAM G4 polar groups and the DPPC head groups. The reduction of ΔH is attributed to the disorganization of the mixture caused by the increase of the incorporated PAMAM G4.

The weakening of the van der Waals forces between the alkyl chains, due to the incorporated PAMAM G4 molecules, gives rise to more fluid bilayers. Since these weak van der Waals interactions dictate the structure of the membrane, we can conclude that at high PAMAM G4 concentrations (very low ΔH values) the lipid bilayer becomes amorphous. The maximum percentage of PAMAM G4 dendrimer that can be incorporated in the DPPC bilayers is 5%, as shown in the thermogram (Fig. 12-2).

Raman spectroscopy helped to specify the exact location of the compound in the lipid bilayer. Experiments at all concentrations were held at 25°C in order to assess the change of the interaction between DPPC and PAMAM G4 at increasing concentrations. The intensity ratios $I_{2935/2880}$, and $I_{1090/1130}$ provided us with information about the bending of the alkyl chain and the final methyl group of DPPC. The spectral area 2800-3100cm^{-1} gives us information about the intramolecular interactions between the alkyl chains of DPPC and thus about their conformation. The intensities and the frequencies of that area are susceptive to the changes that occur to the conformation of the lipoid chains due to the transition of the solid to the liquid-crystalline phase.

Table 12-1. Calorimetric parameters.

Sample (x = mol %)	T_{onset} (°C)	SD	T_m (°C)	SD	$T_{1/2}$ (°C)	SD	ΔH (J/g)	SD
DPPC	41,12	0,01	41,43	0,02	41,51	0,02	48,51	0,07
DPPC/PAMAM G4 (x = 3)	39,36	0,01	40,49	0,03	40,61	0,02	18,75	0,9
DPPC/PAMAM G4 (x = 5)	40,02	0,02	40,58	0,02	40,57	0,01	15,53	0,1
DPPC/PAMAM G4 (x = 10)	39,90	0,01	41,50	0,06	41,51	0,05	8,57	0,05
DPPC/PAMAM G4 (x = 15)	40,42	0,01	41,54	0,02	41,54	0,02	6,68	0,13
DPPC/PAMAM G4 (x = 20)	39,93	0,01	40,43	0,02	40,63	0,02	3,15	0,01
DPPC/PAMAM G4 (x = 30)	39,95	0,03	40,39	0,03	40,50	0,03	3,08	0,06

T_{onset}: temperature at which the thermal effect starts; T_m: temperature at which heat capacity (ΔC_p) at constant pressure is maximum; $T_{1/2}$: temperature at which the transition is half completed; ΔH: transition enthalpy.

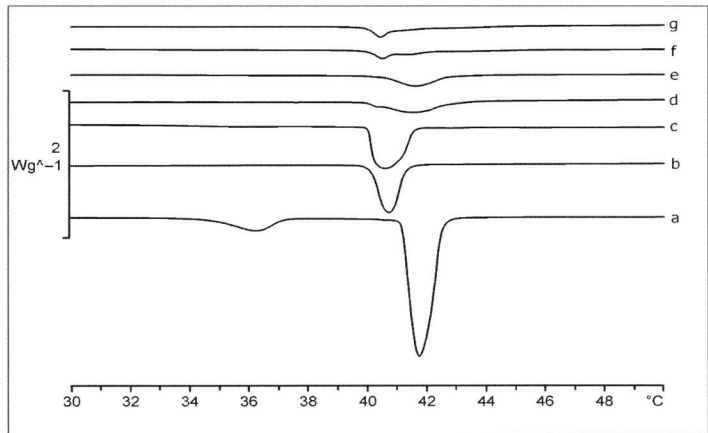

Figure 12-2. DSC thermograms of fully hydrated bilayers of DPPC with varying amounts (a: 0, b: 3, c: 5, d: 10, e: 15, f: 20, g: 30 mol%) of PAMAM G4 dendrimer.

    The incorporation of molecules, such as dendrimers, into the lipid bilayers provokes changes of their conformation and, as a consequence, of the intensity and frequencies of this area. The peaks we studied appear at $2844cm^{-1}$ and $2880cm^{-1}$ and are attributed to the symmetrical and non-symmetrical vibration of bending of the C-H bond of the methylene groups and the peak at $2935cm^{-1}$ is attributed to the symmetrical vibration of bending of the C-H bond of the final methyl group of the alkyl chain. The intensity ratio of the peaks at $2935cm^{-1}$ and $2880cm^{-1}$ is representative of the interaction between the alkyl chains and of their conformation. More specifically this ratio can be used as an index that demonstrates the change of the proportion between disorder and order that exists in the conformation of the alkyl chain, as a factor of temperature. In addition, the area 1000-$1200cm^{-1}$ includes the bending vibrations of the C-C bonds of the alkyl chains of the phospholipids. The peak at $1130cm^{-1}$ is attributed to the bending vibration of the C-C bond for the *trans* conformations of the alkyl chains, while the peak at $1090cm^{-1}$ is attributed to the bending vibration of the C-C bond for the *gauche* conformations of the alkyl chains. The intensity ratio of these peaks can also give us information about the proportion between disorder and order that exists in the conformation of the alkyl chain.

Table 12-2. The intensity ratios $I_{2935/2880}$ and $I_{1090/1130}$ corresponding to the bending degree of the alkyl chain and of the final methyl group of DPPC lipid bilayers at 25°C.

Sample (x = mol %)	$I_{2935/2880}$	$I_{1090/1130}$
DPPC	0,48	0,76
DPPC/PAMAM G4 (x = 3)	0,56	0,82
DPPC/PAMAM G4 (x = 5)	0,71	0,93
DPPC/PAMAM G4 (x = 10)	1,10	0,99
DPPC/PAMAM G4 (x = 15)	0,95	0,99
DPPC/PAMAM G4 (x = 20)	1,33	1,00
DPPC/PAMAM G4 (x = 30)	1,25	1,08

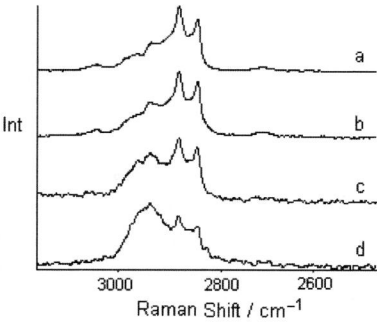

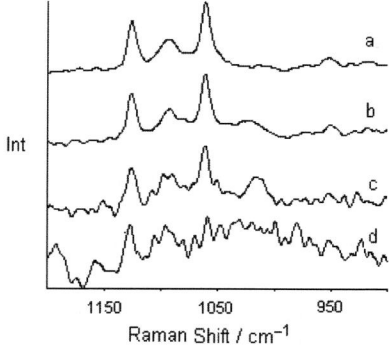

Figure 12-3. RAMAN spectra of DPPC with PAMAM G4 at indicative concentrations of 0%, 3%, 5% and 20% at 25°C and at the spectral areas 2500-3150 cm^{-1} and 920-1200 cm^{-1}. a: DPPC/PAMAM G4 0%; b: DPPC/PAMAM G4 3%; c: DPPC/PAMAM G4 5%; and d: DPPC/PAMAM G4 20%.

As mentioned before, experiments at all concentrations were held at 25°C. The results of the intensity ratios $I_{2935/2880}$ and $I_{1090/1130}$ appear in Table 12-2. An increase in the bending of the carbon-chain and the final methyl group of DPPC was noticed at all incorporated concentrations as concluded by the increase of the intensity ratio $I_{2935/2880}$ and the decrease of the intensity ratio $I_{1090/1130}$. These *gauche/trans* ratios that can be interpreted as disorder/order ratio demonstrate that at a 3% concentration the alkyl chain remains organized and that above 10% the *gauche* conformation is more intense, so the alkyl chain bends leading to its liquidation and its degradation (Fig. 12-3).

5.   CONCLUSIONS

As a conclusion, our study showed that the incorporation of PAMAM G4 dendrimer in DPPC bilayers causes a) abolition of the pre-transition peak of DPPC and b) concentration dependent lowering of ΔH. The intensity ratios and the changes of frequencies of the spectra areas which have been studied by RAMAN spectroscopy claim that the conformational properties of the alkyl chains have been changed due to the incorporation of PAMAM G4 dendrimer into the lipid bilayers. From these facts we can conclude that the physical methods used give us information regarding the interactions between PAMAM G4 dendrimer and DPPC as well as its exact location into the lipid bilayers. These results may justify the tendency of dendrimers to intervene with biological membranes and assess their biocompatibility. The information from this study could prove useful to the rational design of new effective drug carriers consisting of liposomes and dendrimers as a combination of two technologies, for the delivery of bioactive molecules.

6.   REFERENCES

1. Sahoo S.K., Labhasetwar, V. Drug Discovery Today vol. 8, (24) 2003 (www.drugdiscoverytoday.com).
2. Lasic D.D. Liposomes from Physics to Applications, Elsevier 1993.
3. Allen T.M., Stuart D.D. Liposomes Pharmacokinetics in 'Liposomes Rational Design', Ed. A. S. Janoff, Marcel Dekker, Inc., N. York, Basel, 1999, pp. 63-87.
4. Drummond D.C., Meyer O., Hong K., Kirpotin D.B., Papahadjopoulos D. Optimizing liposomes for delivery of chemotherapeutic agents to solid tumors. Pharmacol Rev 1999, 51, 691-743.
5. Woodle M.C. Sterically stabilized liposome therapeutics. Adv. Drug Deliv. Rev. 1995, 16, 249-265.
6. Ferrari M. Cancer nanotechnology: opportunities and challenges. Nature reviews (cancer), 2005, 5, 161-171.

7.   Cloninger M. Biological application of dendrimers. Current Opinion in Chemical Biology 2002, 6, 742-748.
8.   Aulenta F., Hayes,W., Rannard S. Dendrimers a new class of nanoscopic containers and delivery devices. European Polymer Journal 2003, 39, 1741-1771.
9.   Eichman J., Bielinska A., Kukowska-Latallo J., Donovan B., Baker J., in Frechet J. and Tomalia D. *(eds.)* Dendrimers and other Demdritic Polymers J. Wiley & Sons Chisester 2001, pp. 441-462.
10. Khuloud A.l-Jamal, Sakthivel T., Florence A. Dendrisomes: cationic lipidic dendron vesicular assemblies. International Journal of Pharmaceutics 2003, 254, 33-36.
11. Tomalia D.A., Dewald J.R., Hall M.R., Martin S.J., Smith P.B., 1st SPSJ International Polymer Conference Society of Polymer Science Japan, Kyoto, 1984, 65.
12. Tomalia D.A., Frechet J.M.J. Discovery of dendrimers and dendritic polymers: A brief historical perspective. J. of Polymer Science: Part A: Polymer chemistry 2002, 40, 2719-2728.
13. Tomalia D.A., Birth of a new macromolecular architecture: dendrimers as quantized building blocks for nanoscale synthetic organic chemistry, Aldrichimica Acta, vol. 37, No. 2, 2004, 45-46.
14. Tomalia D.A., Frechet J.M.J., Dendrimers and other dendritic polymers, John Wiley and Sons, Ltd, 2001.
15. Georgiadis G., Allison A., Liposomes in biological systems, John Wiley and sons, 1980.
16. Ford J., Timmins P., Pharmaceutical thermal analysis, techniques and applications. (1989), Ellis Horwood unlimited.
17. Koynova R., Caffrey M., Phases and phase transitions of the phosphatidylcholines Biochem.Biophys. Acta 1998, 1376, 91-145.
18. Golthup N.B., Daly L.H., Wiberley S.E., Introduction to Infrared and Raman Spectroscopy, Third Edition 1990, Academic Press, Inc., Harcourt Brace and Company, Publishers, 1250 Sixth Avenue, San Diego, CA 92101.

Index